IRCK: THE FIRST 6 YEARS

The International Research Center on Karst (IRCK) Under the Auspices of UNESCO
Institute of Karst Geology, Chinese Academy of Geological Sciences
Key Laboratory of Karst Dynamics, MLR and GZAR

SCIENCE PRESS
BEIJING

国际岩溶研究中心 6 年历程

联合国教科文组织国际岩溶研究中心
中国地质科学院岩溶地质研究所　编著
国土资源部 / 广西岩溶动力学重点实验室

科学出版社
北　京

内 容 简 介

2013年11月25、26日，由联合国教科文组织聘请组成的专家评估组对国际岩溶研究中心第一个6年工作进行评估。专家评估组认为：国际岩溶研究中心已经成为一个高效的教科文二类研究中心；作为一座桥梁，促进了国际岩溶知识的交流与应用，扩大了岩溶知识在全球范围的影响；用突出的成绩完成了协议规定的任务。本图册是在评估过后，依托大量的照片，反映国际岩溶研究中心在第一个6年建设期的历程，以及国际岩溶研究中心、与岩溶相关的国际组织、相关的国内外岩溶专家对中心成立、建设的支持。图册共分八个章节：中心成立与背景、组织建设与管理、科学研究、学术会议、交流与合作、科普与咨询、国际培训和总结。

为了方便中外相关读者阅读，本图册采用中英文对照，以照片、图表为主，文字简明扼要，适合从事地学、生态学及相关中心、重点实验室管理等方面的科研人员、管理人员阅读、参考。

图书在版编目(CIP)数据

国际岩溶研究中心6年历程 / 联合国教科文组织国际岩溶研究中心，中国地质科学院岩溶地质研究所，国土资源部/广西岩溶动力学重点实验室编著. —北京：科学出版社，2015.1
 ISBN 978-7-03-042174-6

Ⅰ. ①国… Ⅱ. ①联… ②中… ③国… Ⅲ. ①岩溶－研究中心－国际组织－概况 Ⅳ. ①P642.25-20

中国版本图书馆CIP数据核字(2014)第240578号

责任编辑：罗 吉 许 瑞 / 责任校对：蒋 萍
责任印制：李 利 / 封面设计：许 瑞

科学出版社 出版
北京东黄城根北街16号
邮政编码：100717
http://www.sciencep.com

中国科学院印刷厂印刷
科学出版社发行 各地新华书店经销

*

2015年1月第 一 版 开本：889×1194 1/16
2015年1月第一次印刷 印张：25 1/4
字数：872 000

定价：598.00元
（如有印装质量问题，我社负责调换）

IRCK: THE FIRST 6 YEARS Editorial Board

Editor: Jiang Yuchi
Subeditor: Cao Jianhua
Technical Adviser: Yuan Daoxian
Members of Editorial Board: (Sequencing in Pinyin of name order)
Erin Lynch (USA) Chen Hongfeng Chen Weihai Deng Zhaoyi Gan Fuping Guo Fang
He Shiyi Huang Fen Jiang Guanghui Jiang Zhongcheng Lei Mingtang Li Jin Li Qiang
Liu Shaohua Liu Zaihua Lu Qian Luo Qukan Luo Weiqun Miao Ying Ning Duihu
Ou Yiru Pan Moucheng Pei Jianguo Pu Junbing Su Luxuan Sun Pingan Tang Wei
Tang Qingjia Wang Jinliang Wu Xia Xiao Qiong Xie Yincai Xie Yunqiu Yang Hui
Yang Chuchang Yang Lichao Yi Lianxing Yu Shuang Zhang Cheng Zhang Qiang
Zhang Chunlai Zhang Fawang Zhang Jinsong Zhang Meiliang Zhang Xiaoning
Zhang Yuanhai Zhong Kaiwei Zhu Mingqiu Zhu Xiaoyan

《国际岩溶研究中心 6 年历程》编委会

主　　编：姜玉池

副 主 编：曹建华

技术顾问：袁道先

编　　委：（按姓名汉语拼音排序）

Erin Lynch (USA)	陈宏峰	陈伟海	邓朝义	甘伏平
郭　芳	何师意	黄　芬	姜光辉	蒋忠诚	雷明堂
李　晋	李　强	刘绍华	刘再华	卢　茜	罗劢侃
罗为群	苗　迎	宁堆虎	区绎如	潘谋成	裴建国
蒲俊兵	苏橹萱	孙平安	唐　伟	汤庆佳	汪进良
吴　夏	肖　琼	谢银财	谢运球	杨　慧	杨初长
杨利超	易连兴	于　奭	章　程	张　强	张春来
张发旺	张劲松	张美良	张小宁	张远海	钟开威
朱明秋	朱晓燕				

Foreword

The International Research Center on Karst (hereinafter referred to as "IRCK") under the Auspices of the United Nations Educational, Scientific and Cultural Organization (hereinafter referred to as "UNESCO") has successfully passed the first six-year review by the Experts Panel of UNESCO, indicating that it has made a good start towards its goal.

On February 11, 2008, an Agreement Between the Government of the People's Republic of China and the United Nations Educational, Scientific and Cultural Organization Concerning the Establishment and Operation of the International Research Center on Karst(IRCK) in Guilin, China under the Auspices of UNESCO was signed between the Government of the People's Republic of China and UNESCO in Paris, France. On December 15 that year, IRCK was formally launched and started its operation in Guilin, China. IRCK is the first Category II center under the Auspices of UNESCO in geosciences. Based in Guilin, China, IRCK cannot have been made possible without the innovative efforts of Chinese karstologists and the strong support from Chinese Government. The establishment of IRCK also proves to be a milestone in the field of global karst research.

Since the inception of IRCK, the Chinese Government has been giving top priority to the construction, organization and operation of IRCK. In particular, the government agencies including the Ministry of Land and Resources, together with the China Geological Survey, the Chinese Academy of Geological Sciences, and the Institute of Karst Geology, among others, have provided supports in terms of human resources, site, equipment and organization necessary for the building and operation of IRCK at various levels.

Over the past six years, all the members of IRCK worked very hard and achieved a series of outstanding results:
1. Implementing scientific research in an innovative manner: IRCK

has focused on three major themes (i.e. karst dynamic system and global climate change; karst hydro-geological survey, water resources assessment, development and use; karst dynamic system and sustainable development) to conduct scientific research. In particular, it has integrated karst research with global climate change, an international priority topic, by paying special attention to the studies on karst process and global carbon cycle, and the high-resolution records of past climate change from cave stalagmites. The research findings on karst carbon process and effect in China were published with a title of "An unsung carbon sink" in Science (Volume 334) on 18 November, 2011. In September 2013, the Ministry of Science and Technology approved the foundation of "International Joint Research Center on Karst Dynamic System and Global Change" at national level, and the China Geological Survey endorsed the establishment of "Business Center on Geological Research Addressing Global Climate Change", both of which were based on IRCK and the Institute of Karst Geology.

2. Extensively conducting international exchange and initially putting in place an information sharing platform: IRCK has conducted extensive academic exchange within international karst research community to create a first-rate research and information sharing platform. (1) IRCK has maintained close links with the members of the Governing Board and Academic Committee, updating them with the progress of IRCK through quarterly and annual reports; until now, four meetings of the Governing Board and three meetings of the Academic Committee have been held. (2) IRCK has organized, co-organized or participated in a number of international conferences on karst research. Specifically, it has organized and co-organized three international symposia on karst research. For instance, the International Symposium of Karst Water under Global Change Pressure, sponsored by the Karst Committee under IAH and the Chinese Geological Survey and organized by IRCK/ the Institute of Karst Geology, was convened in Guilin, China in April 2013, bringing together

138 scientists and experts from 13 countries worldwide. In addition, the IRCK delegation have taken part in 22 international conferences on karst research. (3) It has kept close contact and established partnership with relevant international organizations. Members of IRCK have attended three meetings of IGCP Science Committee, two events of International Hydrological Programme (IHP), three meetings under International Association of Hydrogeologists (IAH); co-organized a special session on Karst under International Geological Congress (IGC); joined one international conference of International Geographical Union (IGU), and one international conference of Information Systems Committee (ISC); and published a Special Edition on Karst Rocky Desertification jointly with Man and the Biosphere (MAB) National Committee of China. (4) IRCK has developed a network on international collaboration, visited other international centers on karst research to conduct exchanges on management experiences for four times, received visitors 22 times, and signed 13 Memorandum of Understanding (MOU). All these efforts have laid a solid foundation for next steps of IRCK.

3. Undertaking international training programs: IRCK has successfully organized five training courses, which were attended by a total of 87 trainees from 24 countries and 65 facilitators from 17 countries. The topics covered all the areas related to karst dynamic system, and the format of training included classroom learning, field practice and trainee's assessment. Members of the Academic Committee of IRCK have played a crucial role in the process. Remarkable achievements have been made in these training courses. Many trainees, after having attended the first training courses, were also actively engaged in the second or third training courses. The trainees and facilitators have communicated and interacted frequently. Some trainees went back to the training workshop as trainers, making the training workshop become a platform which brings international karstologists together to discuss topics on karst research. Many trainees returned to

their home countries to conduct research projects with new knowledge. For instance, the trainees from Thailand, after having participated in the training courses, led a delegation to visit IRCK, and successfully applied a karst geological survey and research project at national level, which has significantly promoted research effort on karst dynamic system in Thailand.

All in all, I am deeply impressed with the great progress that IRCK has made in the first six years.

I would like to send my congratulations to the publication of *IRCK: The First 6 Years* with the comment by the Experts Panel of UNESCO: as a bridge, IRCK, as expected by the Natural Sciences Sector of UNESCO, has promoted the exchange and application of karst knowledge, expanded the influence of karst knowledge across the world, and effectively implemented its tenet with outstanding achievements.

Finally, I hope that IRCK can better play its vital role in leading the research on karst, continue to work in an innovative manner, contribute to the international cooperation on karst carbon cycle and climate change research, and support the sustainable development of karst areas in the future.

Vice Minister of the Ministry of Land and Resources of the People's Republic of China

Chair of the Governing Board of IRCK

September 2014

序

联合国教科文组织（以下简称教科文组织）赞助的国际岩溶研究中心（以下简称岩溶中心），顺利通过联合国教科文组织专家组第一个6年评估，这意味着岩溶中心取得良好的开端，迈出了坚实的一步。

2008年2月11日，中华人民共和国政府与教科文组织在法国巴黎签署了《中华人民共和国政府与联合国教育、科学及文化组织关于在中国桂林建立由教科文组织赞助的国际岩溶研究中心及其运作的协定》。2008年12月15日，教科文组织岩溶中心在中国桂林挂牌成立并正式运行。岩溶中心是教科文组织在地学领域的第一个二类中心，她落户中国桂林，与中国岩溶地质工作者长期的创新性工作、中国政府的大力支持密切相关，是中国岩溶研究、全球岩溶学术界具有里程碑意义的大事件。

岩溶中心成立以来，我国政府高度重视中心的筹建、组织建设和运行，国土资源部等有关部门以及中国地质调查局、中国地质科学院、岩溶地质研究所等单位分别在不同层面为岩溶中心的建设和运行提供了必要的人力资源场所、设备和组织保障。

国际岩溶研究中心全体成员，不辱使命，不断学习，在实践中探索，在探索中进步，不断总结经验，取得一系列可喜的成果。

创新性地开展科学研究：重点围绕岩溶动力系统与全球气候变化、岩溶水文地质调查与水资源评价和开发利用、岩溶动力系统与可持续发展3个方面，开展科学技术研究。尤其是通过岩溶作用与全球碳循环、洞穴石笋对过去气候变化的高分辨率记录2方面，将岩溶研究与应对全球气候变化这一国际优先主题联系起来，创新性地工作。其中，中国岩溶碳汇过程与效应的研究进展已以题为"An unsung carbon sink"在美国《科学》杂志第334卷（2011年11月18日）发表。

2013年9月，中国科技部批准了国家级"岩溶动力系统与全球变化国际联合研究中心"、中国地质调查局成立"应对全球气候变化地质研究业务中心"，均依托岩溶中心/岩溶所运行。

广泛地国际交流、初步建成信息共享平台：放眼全球，广泛交流，积极主动工作，打造一流研究与信息共享平台：（1）与理事会理事、学术委员会委员保持密切联系，通过季报、年报，就岩溶中心的进展情况及时沟通交流；成功地举办4次理事会、3次学术委员会。（2）承办、协办和参加相关国际会议，举办国际学术研讨会3次、协办3次。如2013年4月，由国际水文地质学家协会岩溶专业委员会、中国地质调查局主办、岩溶中心/岩溶所承办的"岩溶资源、环境与全球变化——认识、缓解与应对"国际研讨会在桂林召开，来自13个国家和地区的138名专家、学者参加会议，取得很好的效果；参加岩溶相关的国际学术会议22批次。（3）与相关国际组织取得联系、建立合作关系，参加IGCP科学委员会3次，参加IHP的相关活动2次，参加IAH国际会议3次，协办IGC大会岩溶专题会，参加IGU国际会议1次、ISC国际会议1次，与MAB中国国家委员会联合出版"岩溶石漠化专辑"。（4）构建国际合作网络，拜访与岩溶相关的国际研究中心进行管理经验交流4批次，学术接待22批次，签署合作备忘录13份。

这些工作为下一步岩溶中心的建设和发展打下了坚实的基础。

国际培训取得成效：成功举办5次国际培训班，吸引了6大洲24个国家87名学员参与培训，邀请17个国家65名岩溶专家、学者作为教员参加培训班，培训内容包括岩溶动力系统研究的各个领域，培训形式包括课堂讲座、野外实践、学员评估等环节。培训期间，岩溶中心学术委员会成员发挥了不可替代的积极作用。

培训班取得很好的效果，很多学员参加第一期培训班后，又积极争取参加第二期、第三期培训。学员与教员之间交流、互动频繁，多个学员回国后，通过一阶段工作，又作为教员，来到培训班进行交流，使得培训班成为每年一度的国际岩溶学者汇聚的平台。很多学员培训后，带着新知识、新认识，回国开展了卓有成效的工作，如泰国学员参加培训后，带领访问团再次访问岩溶中心，并成功申请获批国家级岩溶地质调查研究项目，极大地推动了泰国岩溶动力系统研究。

我很高兴看到岩溶中心在第一个6年建设期取得的成效与成果。

最后，我想用教科文组织专家评估组的意见来表达对《国际岩溶中心6年历程》正式出版的祝贺：岩溶中心按照教科文组织自然科学部的期望那样，作为桥梁，促进了岩溶知识的交流与应用，扩大了岩溶知识在全球范围的影响，用突出的成绩有效履行了它的宗旨。

希望岩溶中心能更好地发挥在岩溶领域的引领带动、桥梁纽带作用，继续创新性地努力工作，促进岩溶碳循环和气候变化国际合作，促进和支持岩溶地区可持续发展。

国土资源部副部长
国际岩溶研究中心理事会主席

2014 年 9 月

本书照片来源广泛，故不一一署名，对本书有贡献的人名统一放在编委名单中。

国际岩溶研究中心 6 年历程 目录
CONTENTS

Chapter 1　Establishment and Background

| 002　第一章　中心成立与背景

Chapter 2　Organization and Management

| 024　第二章　组织建设与管理

Chapter 3　Scientific Research

| 058　第三章　科学研究

Chapter 4 Academic Meetings

164　第四章　学术会议

Chapter 5 Exchange and Cooperation

206　第五章　交流与合作

Chapter 6 Science Dissemination and Consulting Service

272　第六章　科普与咨询

Chapter 7　International Training

306　第七章　国际培训

Chapter 8　Summary

362　第八章　总结

Fengcong karst is defined as a group of rocky hills or peaks rising from shared limestone foot-slopes. Closed depressions lie between the peaks, so the landscape is sometimes described as peak cluster depression. Fenglin karst is defined as limestone towers or peaks that are isolated from each other by flat limestone surfaces, which are generally covered by a thin layer of loose sediment. The peaks are usually completely surrounded by a karst plain, so the landscape may be called a peak forest plain.

Guilin features world-class fenglin-fengcong karst landscape.

峰丛地形指具有连座的一些石峰，其间常有封闭洼地，因此其组合形态也可称为峰丛洼地。狭义的峰林地形是指被一片平地所分割的一些石峰，其组合形态可以分为峰林平原、峰林谷地、峰林盆地等。桂林是岩溶峰林、峰丛最为典型的代表。

（本页照片由朱学稳提供）/ This photo is from Zhu Xuewen

Chapter 1
Establishment and Background

第一章　中心成立与背景

联合国教科文组织国际岩溶研究中心在中国桂林成立，该中心是联合国教科文组织赞助设立的第一个地学二类研究中心，也是在国际地学计划（IGCP）框架下成立的第一个二类研究中心。

国际岩溶研究中心的成立主要有以下几个方面的支撑：

(1) 岩溶在地球表面具有广泛的分布，其分布面积占陆地面积的10%–15%，近1/4的人口依赖岩溶水生产、生活；

(2) 由碳酸盐岩形成的岩溶环境是脆弱的生态环境，碳酸盐岩沉积在清洁的海洋环境中，因此碳酸盐岩风化溶解可提供的成土物质先天不足，导致土壤资源短缺；岩溶含水层的双层结构使岩溶区地表水资源短缺、地下水资源丰富，这不仅给水资源的开发利用带来较大困难，同时也导致岩溶区，尤其峰丛洼地区的旱涝灾害频繁发生；

(3) 在岩溶环境中，人类生产、生活的水土资源短缺，发展受到制约，因此，分布在岩溶区人口相对贫困；

(4) IGCP项目的执行，不仅形成全球范围岩溶科学家联系平台，而且为中心建设的目标提供科学基础；岩溶动力系统理论成为中心的理论基础；

(5) 国际岩溶研究中心的成立还得到与岩溶相关的国际学术组织和相关国家的支持，得到中国政府的大力支持。

国际岩溶研究中心成立在教科文组织前期考察、评估的基础之上，国土资源部代表中国政府，于2008年2月在联合国教科文组织巴黎总部签署，2008年12月正式挂牌成立。

这是对中国岩溶研究进展的肯定，同时在岩溶学术界也是具有里程碑意义的大事件。

The International Research Center on Karst (IRCK) under the Auspices of UNESCO was established in Guilin, China. It is the first category II center concerning geosciences under the auspices of UNESCO and in the framework of the International Geoscience Programme (IGCP).

IRCK's establishment was motivated by a number of key factors:

(1) Karst comprises 10%–15% of the Earth's land surface area and nearly a quarter of the world's population relies on karst water.

(2) The karst environment is fragile. Weathering and dissolution of carbonate rocks leaves little residual material for soil formation, resulting in a shortage of soil. Due to the double-layer structure of karst aquifers, karst areas suffer from a shortage of surface water, while groundwater is abundant. The nature of karst causes difficulty in development and utilization of water resources. Additionally, karst areas are frequently susceptible to drought and flood disasters, especially in areas of peak cluster depressions.

(3) A shortage of water and soil resources in karst areas restricts development, resulting in poverty.

(4) The IGCP karst projects produce a communication platform for karst scientists around the world, and provided a scientific basis for the construction of the center. The karst dynamic system (theory) is the theoretical basis for the center.

(5) The establishment of IRCK was supported by international academic organizations, national karst institutes, and the Chinese Government.

IRCK was established following evaluation by UNESCO. In February 2008, a representative of the Chinese Government from China's Ministry of Land and Resources signed an agreement at UNESCO's Paris headquarter. IRCK was formally founded in December 2008.

IRCK's establishment is recognized as the progress of Chinese karst research, and also a landmark event for international karst academia.

Karst Covering a Wide Area, constituting 10%–15% of the Earth's Total Land Surface Area
全球岩溶分布面积广，占陆地面积的 10%–15%

Carbonate rocks are predominately from the Mesozoic (Jurassic–Cretaceous) (39.1%), Lower Paleozoic (19.5%), Cenozoic (15.4%), Precambrian (11.2%) and Upper Paleozoic (10.2%). Over 60% of the carbonate rocks are located between 20°–50°N worldwide, carbonate outcrops account for a significant percentage of the continental land area: Asia (9.57%), Africa (10.97%), North America (15.31%), Europe (19.95%), South America (1.90%) and Australia (6.78%).

在时间上，碳酸盐岩地层主要来自于中生代（侏罗纪—白垩纪）(39.1%)、下古生代 (19.5%)、新生代 (15.4%)、前寒武系 (11.2%) 和上古生代 (10.2%)。

在空间上，超过 60% 的碳酸盐岩分布在 20°–50°N，亚洲、非洲、北美洲、欧洲、南美洲、大洋洲的碳酸盐岩出露的面积占陆地面积的比例分别为 9.57%、10.97%、15.31%、19.95%、1.90%、6.78%。

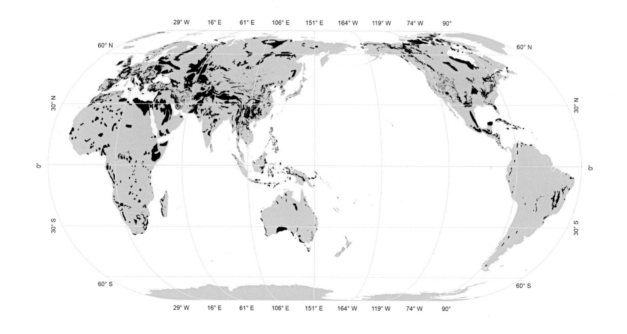

Global distribution of karst
(Karst areas indicated in black, Willians and Ford, 2006)
全球碳酸盐岩的空间分布

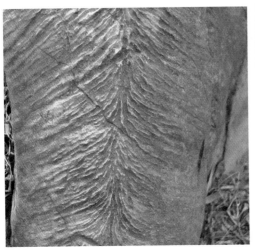 

上左图
Top left
Limestone forms regular weathering surfaces.
石灰岩的风化表面因存在龙脊出现规则排列的羽状溶纹。

上右图
Top right
Dolomite forms an irregular weathering surface, with dissolution occurring along multipe axes.
白云岩的风化表面呈不规则的交错条纹状。

Carbonate area by continent (Hans et al., 2005)
碳酸盐岩在各大洲分布的面积及占陆地面积比例统计表（Hans et al., 2005）

	Asia 亚洲	Africa 非洲	North America 北美洲	Europe 欧洲	South America 南美洲	Australia 大洋洲	Total 合计
Karst area / ($\times 10^6 \text{km}^2$) 面积 / ($\times 10^6 \text{km}^2$)	4.20	3.30	3.41	1.9	0.34	0.61	13.82
Percentage of worldwide carbonate rocks per contnient / % 各大洲碳酸盐岩面积占总碳酸盐岩面积的比例 / %	30.39	23.88	24.67	14.18	2.46	4.41	100
Percentage of continental land area occupied by carbonate rocks / % 碳酸盐岩面积在各大洲的比例 / %	9.57	10.97	15.31	19.95	1.90	6.78	—

The Fragile Karst Ecosystem: Scarce Water and Soil Resources
岩溶生态环境是脆弱的生态环境：缺水、少土

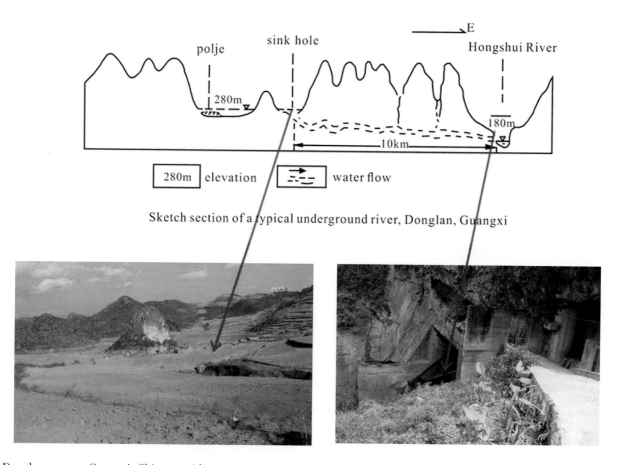

Donglan county, Guangxi, China providing an example of the double-layer hydrogeological structure found in karst areas
广西壮族自治区东兰县岩溶水文地质结构的双层性示意图

Here the local population lives in the recharge area of the underground river system: an area of depressions and poljes. Rainfall in the recharge area sinks quickly into the underground river system, and discharges to the surface at the downstream resurgence.

洼地、坡立谷是地下河的上游补给区，同时也是当地居民主要的生产、生活所在地。洼地、坡立谷中的降水沿漏斗、落水洞快速进入地下，在下游排泄进入地表河流。

In karst region there is little soil, and residents must resort to farming amidst the rocky outcrops.
由于岩溶区土壤稀少，只能开垦、耕种石旮旯地成为当地居民的基本口粮田。

In Dahua County, Guangxi tectonic uplift has made it possible to clearly see the double-layer hydrogeological structure in the Qibainong peak cluster depression karst area. The plateau has depressions and dolines, suffering from soil scarcity as well. Water flowing from the upper level into the underground river system transported soil, which aggravate the soil erosion and water shortage on the plateau.

广西大化县七百弄峰丛洼地区因构造运动，地下河出现巨大的天窗，可以清晰看到岩溶区双层的水文地质结构，地表是洼地、漏斗，地下是地下河，同时地表土壤资源稀少，大量土壤随水流进入地下河，更加重地表水土资源短缺。

Karst and Poverty in Southwest China
中国西南岩溶区人口贫困

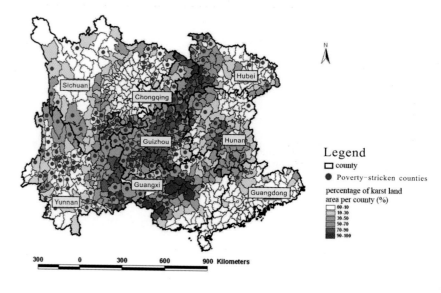

Karst and poverty by county in southwest China
中国西南岩溶区岩溶县分布与国家贫困县空间分布的对应关系

In this figure karst areas are colored, with darker colors indicating a higher percentage of karst land area per county. Red dots indicate counties which appear on China's list of national-level poverty-stricken counties. The fragile karst ecological environment restricts the economic and social development in poverty-stricken areas.

图中有颜色的是岩溶县，颜色越深，表示该县岩溶分布面积占土地面积的比例越高，红点代表国家贫困县。脆弱的生态环境、贫困的人口制约着经济社会的发展。

第一章　中心成立与背景　011

A village, karst rocky desertification in Liupanshui city, Guizhou, China
贵州省六盘水市岩溶石漠化区自然村落

Here the hills behind the village are denuded, with sparse scrubs growing on exposed limestone. The flat area in front of the village also has exposed limestone and the soil is very thin. Comprehensive control of rocky desertification of karst areas, improving the environment for the residents, and increasing the residents' income, are the aims of the Chinese Government in the period of "11th Five-Year Plan" (2006-2010) and "12th Five-Year Plan" (2011-2015) in karst area.

村后裸露的石灰岩上生长着零星的灌草，村前是稍有平整的耕地，耕地中可见出露的石灰岩，土层非常薄。开展岩溶区石漠化综合治理、改善当地居民的生存环境、提高当地居民的收入、脱贫，是中国政府在"十一五"、"十二五"间实施的"岩溶区石漠化综合治理工程"的目标。

对页左上图
Opposite page, top left

From 2005 to 2006, China Ministry of Water Resources organized a survey entitled "comprehensive survey on water and soil loss and ecological security", in Du'an County in western Guangxi. This photo captures the difficulties of living and farming in karst areas. The woman (left) is carrying a basket of sweet potatoes from farmland which is far from her village; a woman (right) is carrying a tank to take water from an underground river.

本照片是在2005—2006年水利部组织的"全国水土流失与生态安全综合考察"时，在广西西部都安县抓拍到的当地的艰难生产、生活情景。最左面是一位妇女从山外背回一筐红薯，意味着她家的口粮田距她居住的家有较长的距离；最右面是一位妇女背了一个水缸，要到地下河去取水。

对页左下图
Opposite page, bottom left

Mountainous karst regions were severely affected by the 2009-2010 drought in southwest China. five million residents had insufficient drinking water and crop production decreased by 3.5 million tons. During the drought villagers carried drinking water from distant underground rivers (Photo provided by Xinhua).

2009—2010年西南大旱期间，岩溶石山区是干旱最为严重的地区，500万居民缺乏饮用水，粮食减产350万吨。为了解决饮用水问题，所有人员都到很远的地下河区担水。本图为新华网提供的照片。

IGCP Projects Expanded the Field of Karst Research
IGCP 项目的执行拓展了岩溶研究领域

Since 1990, Chinese scientist Prof. Yuan Daoxian has proposed and hosted three successive international karst projects, they were:

(1) IGCP 299 "Geology, Climate, Hydrology and Karst Formation" (1990–1994)

(2) IGCP 379 "Karst Processes and the Global Carbon Cycle" (1995–1999)

(3) IGCP 448 "World Correlation on Karst Geology and Its Relevant Ecosystem" (2000–2004)

These projects offered opportunities to solve the most urgent problems for karst environment and resources, and formed a powerful international cooperation team or group, that strongly drove the development of karst science both internationally and within China.

Through great efforts of karst scientists, the IGCP projects not only increased our knowledge of karst, but also put forward a new concept for karst science – the karst dynamic system.

In addition, two other IGCP projects related to karst were IGCP 513 "Global Study of Karst Aquifers and Water Resources" (2005–2009) and IGCP/SIDA 598 "Environmental Change and Sustainability in Karst Systems" (2011–2015).

1990 年以来，我国科学家袁道先院士连续提出申请并主持实施了三个国际岩溶对比计划项目。这三个项目分别是：

(1) IGCP 299 "地质、气候、水文与岩溶形成"（1990-1994）

(2) IGCP 379 "岩溶作用与碳循环"（1995-1999）

(3) IGCP 448 "岩溶地质及其相关的生态系统全球对比"（2000-2004）

它们为国际岩溶学术界（含 IAH、IGU、UIS 的岩溶委员会）提供了共同解决最紧迫的岩溶资源环境问题的机会，而且形成了一个强有力的国际合作群体，从而有力地推动了国际和我国岩溶学科的发展。

在岩溶科学家共同努力下，项目取得了很多新的认识，其中最为重要的是基于地球系统科学指导下，提出岩溶动力系统理论。

随后与岩溶相关的IGCP项目，分别为 IGCP 513 "岩溶含水层与水资源全球研究"（2005-2009）和 IGCP/SIDA 598 "岩溶系统中的环境变化与可持续性"（2011-2015）。

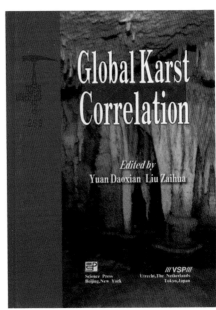

The summary conference for IGCP 299 was held at the University of Oxford in September 1994. At the conference, the chairman of the IGCP Board, renowned geologist Prof. Sir G. Malcolm Brown, FRS praised the IGCP 299 as one of the most productive projects executed, and personally acknowledged its success in UNESCO's "Nature and Resources" publication.

IGCP 299 项目的总结大会于 1994 年 9 月在英国牛津大学举行，在大会上，时任 IGCP 执行局主席、著名地质学家 M.Brown 勋爵（皇家学会会员）致词祝贺，赞扬 IGCP 299 是执行得最富有成果的项目之一，并亲自撰文在 UNESCO 刊物 Nature and Resources 上做了详细介绍。

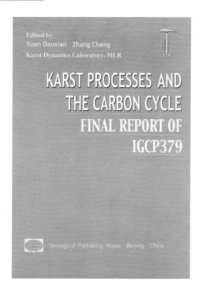

More than 140 individuals from over 40 countries participated in IGCP 379, and monitoring sites were established in Canada, the United States, Russia, Japan, China, Austria, Slovenia, Spain, Poland, and Norway. The project achieved a new understanding of karstification and the carbon cycle through both field and laboratory work.

In-lab simulation of water flow by rapidly spinning submerged rock samples on a turntable at the university of Bremen in Germany revealed that carbonic anhydrase can promote the dissolution rate of limestone and dolomite.

Estimated carbon flux due to carbonate rock dissolution removes CO_2 from the atmosphere at a rate of 1.774×10^7 ton per year in China, and 6.08×10^8 ton per year globally. The global figure accounts for roughly one third of the missing carbon in the global carbon cycle.

IGCP 379 共有40多个国家的140多名代表参加该项目工作，其中加拿大、美国、俄罗斯、日本、中国、奥地利、斯洛文尼亚、西班牙、波兰、挪威等国建立了相应的岩溶作用与碳循环的监测站点，通过大量的野外和室内工作，取得新的认识。

在德国不来梅大学的室内转盘模拟实验室中，揭示了碳酸酐酶对加速石灰岩、白云岩溶解速率的促进。

估算了中国和全球岩溶作用回收大气 CO_2 的量，中国每年因碳酸盐岩溶解消耗大气 CO_2 为 1.774×10^7 tc/a，而全球年回收如按 6.08×10^8 tc/a 计，约占当时碳循环模型中的遗漏汇 (missing sink) 的 1/3。

A joint meeting of IGCP 448's international working group and the "Sustainable Development in Karst Regions" was held on Aug. 30–31, 2001 in Beijing, China. Following the meeting, 38 participants from 12 countries took part in a field excursion covering 4000 km during Sept. 1–12, 2001. The itinerary included Beijing and karst regions in southern China (Chongqing; Liupanshui, Guizhou; Stone Forest, Yunnan and Guilin, Guangxi). The IGCP 448 secretariat provided participants with a 94-page guidebook. The karst researchers who took part in the field seminar visited and compared karst landscape and related karst ecosystem and geological setting in both south and north China.

IGCP 448 国际工作组和 2001 年 8 月 30 – 31 日在中国北京召开的"岩溶地区可持续发展"国际研讨会联合，在会后举行了由来自 12 个国家 38 人参加的行程达 4000 km 的 IGCP 448 野外考察，时间是 2001 年 9 月 1 – 12 日，地点是北京和中国南方岩溶地区（重庆、贵州六盘水、云南石林和广西桂林），IGCP 448 秘书处向野外考察参加者发了 94 页的导游词，用于参观和对比中国南、北岩溶景观，生态环境和形成背景。

IRCK Received Supports from International Karst Organizations and National Karst Institutes
得到与岩溶相关的国际组织和相关国家岩溶机构的支持

上左图 / Top left

IRCK's application was strongly supported by the Chinese Government and administrative departments, especially the Ministry of Land and Resources, China Geological Survey, the Chinese Academy of Geological Sciences, and Guilin City Government. The above figure is showing that, Guilin Mayor Li Jinzao wrote a letter supporting IRCK on February 12, 2002.

国际岩溶研究中心的申请还得到中国政府及管理部门的大力支持，尤其是国土资源部、中国地质调查局、中国地质科学院和桂林市政府。

上左图为 2002 年 2 月 12 日，时任桂林市市长李金早书写的支持信。

上中图 / Top middle

UNESCO's IGCP Scientific Board supported IRCK's application.
In August 2004, IGCP Executive Secretary Wolfgang Eder and Prof. Yuan Daoxian discussed IRCK at the 32th International Geological Congress in Florence, Italy.

国际岩溶研究中心的申请得到联合国教科文组织 IGCP 科学委员会的关照和支持。上中图为 2004 年 8 月，意大利佛罗伦萨第 32 届国际地质大会上 IGCP 秘书 W. Eder 博士与袁道先院士商谈桂林国际岩溶研究中心有关事宜。

上右图 / Top right

With the help of the China National Committee for IGCP, the Permanent Delegation of the PRC to UNESCO, in May 2006 following carried out a field survey and wrote a report. The IGCP Executive Secretary Robert Missotten confirmed that Guilin is a good location for IRCK due to its outstanding karst, the excellent research team and achievements by the IGCP projects led by Prof. Yuan Daoxian.

2006 年 5 月，在 IGCP 中国全委会和中国 UNESCO 常驻团的积极推动下，IGCP 生态与地球科学部秘书 Robert Missotten 博士，通过汇报和野外考察，认为在中国桂林建立国际岩溶研究中心的条件、科研力量及已有的成果基本满足中心建立要求。

The launch of IGCP karst programs extended exchanges and international cooperation, and facilitated many international karst experts for their investigating and furthering understanding of Chinese development karst. When the establishment of the International Research Center on Karst under the Auspices of UNESCO in Guilin, China was proposed, the application received great supports from numerous international karst organizations and national karst institutions, including the International Association of Hydrogeologists (IAH), International Geographical Union (IGU), International Union of Speleology (UIS) and karst research institutions from nearly 20 countries.

通过IGCP项目的执行,广泛的交流和合作,尤其在众多国际岩溶学家考察、认识中国岩溶发育后,在申请中国桂林成立"联合国教科文组织国际岩溶研究中心"提出后,得到与岩溶相关的国际组织和相关国家岩溶机构的支持。如国际水文地质学家学会IAH、国际地理联合会IGU、国际洞穴研究会UIS等岩溶专业委员会,以及来自近20个国家与岩溶相关研究机构的支持。

Signing Ceremony for the Establishment of IRCK
成立国际岩溶研究中心协议的签署

On February 13, 2008, UNESCO's official website reported on the establishment of the International Research Center on Karst under the Auspices of UNESCO.

2008年2月13日，刊登在联合国教科文组织官方网站上的关于在中国桂林成立联合国教科文组织国际岩溶研究中心的新闻报道稿。

第一章　中心成立与背景

Before signing, Mr. Wang Shouxiang, Vice-Minister of China Ministry of Land and Resources, spoke with Prof. Walter Erdelen, UNESCO Assistant Director-General for Natural Science, and Dr. Robert Missotten of UNESCO Division of Ecologial and Earth Sciences.

签字前，王寿祥副部长（左二），同 UNESCO 负责自然科学的助理总干事 Walter Erdelen（右二）及地学生态部主任 Robert Missotten（右一）交谈。

On February 11, 2008, UNESCO Director-General Koichiro Matsuura and Vice Minister Wang Shouxiang of China Ministry of Land and Resources signed an agreement establishing the International Research Center on Karst in Guilin, China as a Category II international research center under the auspices of UNESCO.

Pictured above, the 2nd, 3rd, 4th, 6th and the 7th person from the left side are respectively: Ambassador Shi Shuyun of Permanent Delegation of the PRC to UNESCO, Wang Shouxiang, Koichiro Matsuura, Walter Erdelen, Robert Missotten.

2008 年 2 月 11 日，联合国教科文组织 (UNESCO) 总干事松浦晃一郎先生与中华人民共和国国土资源部副部长王寿祥先生签署了在中国桂林成立国际岩溶研究中心的协议，该中心是 UNESCO 领导下的二类国际研究中心。

左起 2、3、4、6、7：师淑云大使、王寿祥副部长、UNESCO 总干事松浦晃一郎（Koichiro Matsuura）、助理总干事 Walter Erdelen、地学生态部主任 Robert Missotten。

Opening Ceremony for IRCK Establishment in Guilin, China
国际岩溶研究中心落户中国桂林

The opening ceremony was held on December 15, 2008. There were over 200 people to participate, representing more than 20 universities, institutions, and organizations including UNESCO, China Ministry of Land and Resources, Ministry of Education, Ministry of Science and Technology, Ministry of Water Resources, China National Natural Scientific Foundation, Guangxi Regional Government, Guilin City Government, China Geological Survey, National Commission of the PRC for UNESCO, Permanent Delegation of the PRC for UNESCO, and the China National Committee for IGCP.

国际岩溶研究中心于2008年12月15日在中国地质科学院岩溶地质研究所挂牌成立，来自联合国教科文组织、国土资源部、教育部、科技部、国家自然科学基金委员会、水利部、广西壮族自治区政府、桂林市政府、中国地质调查局、中国地质科学院、中国驻联合国教科文组织常驻团等，以及国内外大学、研究机构的代表共200多人参加了成立大会。

IRCK was formally launched on 15 December 2008 by UNESCO Assistant Director-General Walter Erdelen and Vice Minister Wang Min of China Ministry of Land and Resources.

Left to right: Jiang Yuchi, Dong Shuwen, Yuan Daoxian, Chen Zhangliang, Wang Min, Li Zhigang, Walter Erdelen, Fang Maotian, Li Zhijian and Zhong Ziran.

2008年12月15日国际岩溶研究中心成立大会上，助理总干事Walter Erdelen先生及国土资源部副部长汪民先生等为IRCK揭牌。

从左到右：姜玉池、董树文、袁道先、陈章良、汪民、李志刚、Walter Erdelen、方茂天、李志坚、钟自然。

UNESCO Assistant Director-General Walter Erdelen spoke at the opening ceremony and stated that IRCK would be orientated to the development of karst dynamic system, and global dissemination would improve scientific research and technology for sustainable development in karst areas, which was particularly important in the developing countries.

助理总干事 Walter Erdelen 先生在成立大会致辞，期望中心在全球范围推广岩溶动力系统理论，为岩溶区的可持续发展提供科学技术支撑。

Vice Minister Wang Min of China Ministry of Land and Resources expressed his hopes that IRCK would become a good platform for karst research, theoretical innovation, international academic exchange, cooperation and training, consultation and service. He encouraged the IRCK to improve the exploration and development of karst resources, to protect the ecological environment and to provide services for the sustainable development of the whole society of global karst areas.

国土资源部副部长汪民先生在致辞中提出，要将中心建设成一流的国际岩溶研究、理论创新基地；一流的国际岩溶交流、合作、培训和咨询的平台；为全世界岩溶区的经济社会协调、可持续发展提供优质服务。

During their visit to Guilin for the opening ceremony, Prof. Walter Erdelen accompanied with Prof. Dong Shuwen, Secretary-General of China National Committee for IGCP and Prof. Yuan Daoxian visited the Karst Geology Museum of China.

成立大会期间,在IGCP中国全委会秘书长董树文教授、袁道先院士等陪同下,助理总干事Walter Erdelen参观了中国岩溶地质馆。

During the opening ceremony, Mr Chen Zhangliang, vice-chairman of Guangxi Autonomous Region, talked with Mr. Dong Shuwen, Prof. Yuan Daoxian and Prof. Jiang Yuchi, and congratulated the establishment of the IRCK.

成立大会期间,广西壮族自治区副主席陈章良与董树文、袁道先、姜玉池亲切交谈,祝贺国际岩溶研究中心的成立。

Photo caption:
Name: Stone Forest Geopark, World Natural Heritage Site
Location: Shilin Yi Autonomous County in Kunming, Yunnan
Geopark established in 2004
World Natural Heritage Inscription: 2007.06
Summary: The Stone Forest in Shilin, Yunnan is an outstanding example of pinnacle karst. During the late Paleozoic the area was a coastal-neritic environment, and thousands of meters of limestone and dolomite were deposited, laying a foundation for the later formation of the pinnacle karst. Tectonic uplift raised the area above sea level, and dissolution by groundwater and surface water along rock fracture finally formed the numerous forms of pinnacle karst seen there today.

照片说明：
名称：石林地质公园，世界自然遗产
所在地点：云南省昆明市石林彝族自治县
列入地质公园时间：2004年
列入世界遗产时间：2007年6月
概述：云南石林地质公园是一个以石林地貌景观为主的岩溶地质公园。晚古生代这里为滨海浅海环境，沉积了上千米的石灰岩、白云岩，为形成本区石林地貌奠定了基础。经受后期地壳运动的抬升作用成为陆地，多期次遭受地下水、地表水沿岩石裂隙进行溶蚀，最后形成了组合类型多样的石林地貌景观。

（本页照片由刘宏提供）／This photo is from Liu Hong

Chapter 2
Organization and Management

第二章　组织建设与管理

联合国教科文组织国际岩溶研究中心在中国桂林成立，她是联合国教科文组织赞助设立的第一个地学二类国际岩溶研究中心，成立以来在组织结构和管理方面主要做了以下工作，机构方面：(1) 成立第一届理事会，理事会任命第一届中心主任和学术委员会主任；并召开4次理事会；(2) 成立学术委员会，选取学术委员会副主任2人，召开学术委员会3次；(3) 主任任命中心常务副主任、秘书长，开展中心各项工作，成立中心各职能部门，任命相关负责人。管理方面：(1) 形成中心章程，并通过UNESCO的审查；(2)《联合国教科文组织国际岩溶研究中心财务、行政和人员管理办法》；(3)《联合国教科文组织国际岩溶研究中心发展计划（2010-2015年）》；(4) 每季度编制季度报，每年编写年报，并上报UNESCO生态地学部、理事会委员、学术委员会委员；(5) 制作中心网站，通过网站发布相关新闻、信息、招收学员、开展国际培训班；(6) 开展客座研究，加强国际合作。

After the IRCK was founded, the following were carried out to establish the center's organizational structure:

(1) The 1st IRCK Governing Board was formed, the first directors of IRCK and the IRCK Academic Committee were appointed, and four Governing Board meetings were held.

(2) The IRCK Academic Committee was formed, two deputy directors were voted for the Academic Committee, and three Academic Committee meetings were held.

(3) The IRCK Director appointed an executive deputy director and a secretary-general to carry out the work of IRCK, established several functional division, and appointed responsible persons.

The following management tasks were completed:

(1) The IRCK Statutes were written and reviewed by UNESCO.

(2) The IRCK's "Financial, Administrative and Personnel Management Procedures" were passed.

(3) The "IRCK Development Plan (2010-2015)" was made and passed.

(4) Quarterly and annual reports were compiled in a timely fashion and given to UNESCO Division of Ecological and Earth Sciences, Governing Board members, and Academic Committee members.

(5) IRCK created a website for news and information dissemination, student recruitment, and carrying out international training courses, etc.

(6) IRCK conducted the academic exchange with visiting researchers and strengthened international cooperation.

The 1st IRCK Governing Board
国际岩溶研究中心第一届理事会

According to the agreement for IRCK establishment and IRCK statutes, the 1st IRCK Governing Board is comprised of 16 committee members, including 5 foreigners. They are Prof. Wang Min, Dr. Robert Missotten, Zhou Zhaodong, Huang Junhua, Peng Qiming, Zhong Ziran, Dong Shuwen, Zhang Hongren, Dr. Wilhelm Struckmeier, Prof. Chris Groves, Petar T. Milanovic, Prof. Derek C. Ford, Prof. Yuan Daoxian, Prof. Lu Yaoru, Wang Jiyang and Wang Yanxin.

根据协议和中心章程，第一届理事会委员16人，其中中国境外委员5人：汪民、罗伯特·米索腾、周兆东、黄俊华、彭齐鸣、钟自然、董树文、张宏仁、威赫穆·斯图克梅尔、克里斯·格若斯、彼塔·米拉诺维克、德瑞克·福特、袁道先、卢耀如、汪集暘、王焰新。

Governing Board members and distinguished guests took part in the first session of 1st IRCK Governing Board.

Left to right: Hong Tianhua, Wang Yanxin, Lu Yaoru, Yuan Daoxian, Chris Groves, Wang Jiyang, Dong Shuwen, Robert Missotten, Wang Min, Zhang Hongren, Huang Junhua, Zhong Ziran, Zhou Zhaodong, Jiang Yuchi, Li Zhijian.

参加第一届理事会成立大会的委员及嘉宾合影。
从左到右：洪天华、王焰新、卢耀如、袁道先、克里斯·格若斯、汪集暘、董树文、罗伯特·米索腾、汪民、张宏仁、黄俊华、钟自然、周兆东、姜玉池、李志坚。

Chairman of IRCK's 1st Governing Board, Dr. Wang Min, hydrogeologist, and Vice Minister of China Ministry of Land and Resources, Director General of China Geological Survey.

国际岩溶研究中心第一届理事会主席汪民先生，水文地质学家、国土资源部副部长、中国地质调查局局长。

The 1st IRCK Academic Committee
国际岩溶研究中心第一届学术委员会

Top left

IRCK Academic Committee Director Yuan Daoxian, CAS academician, karst scientist, and chairman of the IGCP 299, 379, 448 projects.

国际岩溶研究中心学术委员会主任袁道先生，中国科学院院士、岩溶学家，IGCP 299、379、448 国际工作组主席。

Top middle

IRCK Academic Committee Deputy Director Wang Jiyang, CAS academician, geophysical and geothermal scientist.

国际岩溶研究中心学术委员会副主任汪集暘先生，中国科学院院士、地球物理及地热学家。

Top right

IRCK Academic Committee Deputy Director Ralf Benischke, professor at the Institute of Water Resources Management, Hydrogeology and Geophysics, Joanneum Research, Austria.

国际岩溶研究中心副主任Ralf Benischke先生，奥地利格拉兹岩溶水资源管理与地球物理研究所教授，岩溶水文地质学家。

The Academic Committee members were selected at the recommendation of the IRCK Director and the IRCK Academic Committee Director. The IRCK Academic Committee is comprised of 32 influential karst scientists, including 15 foreigners: Yuan Daoxian, Ralf Benischke, Wang Jiyang, Andrej Kranjc, Andrzej Tyc, Cai Yunlong, Chris Groves, Derek Ford, Francois Zwahlen, Feng Zongwei, Ian Fairchild, Jiang Zhongcheng, Lin Xueyu, Liu Zaihua, Lu Guanghui, Neven Kresic, Nico Goldscheider, Ognjen Bonacci, Pan Genxing, Qi Shihua, Richard Lawrence Edwards, Shen Zhaoli, Tang Changyuan, Tran Tan Van, Wang Jianli, Wang Shijie, Wolfgang Dreybrodt, Xu Yongxin, Xue Yuqun, Yin Yueping, Yu Longjiang and Zhao Jingbo.

国际岩溶研究中心学术委员会委员，经过中心主任和学术委员会主任提议、协商，共邀请32位在岩溶学术界具有较大影响的科学家组成，其中中国境外委员15人：袁道先、Ralf Benischke、汪集暘、Andrej Kranjc、Andrzej Tyc、蔡运龙、Chris Groves、Derek Ford、Francois Zwahlen、冯宗炜、Ian Fairchild、蒋忠诚、林学钰、刘再华、卢光辉、Neven Kresic、Nico Goldscheider、Ognjen Bonacci、潘根兴、祁士华、Richard Lawrence Edwards、沈照理、唐常源、Tran Tan Van、王建力、王世杰、Wolfgang Dreybrodt、徐永新、薛禹群、殷跃平、余龙江、赵景波。

Academic Committee members and distinguished guests took part in the first session of 1st IRCK Academic Committee.

First row: Shen Zhaoli, Richard Lawrence Edwards, Ognjen Bonacci, Lin Xueyu, Derek Ford, Yuan Daoxian, Petar Milanovic, Xue Yuqun, Wang Jiyang, Jiang Yuchi.

Second row: Liu Zaihua, Xu Yongxin, Andrzej Tyc, Ralf Benischke, Chris Groves, Pan Genxing, Jiang Zhongcheng, Wang Shijie, Zhao Jingbo, Tang Changyuan, Lu Guanghui, Zhang Cheng.

Third row: Cao Jianhua, Neven Kresic, Qi Shihua, Andrej Kranjc, Tran Tan Van, Francois Zwahlen, Cai Yunlong, Yu Longjiang, Wang Jianli.

参加第一届学术委员会成立大会的委员及嘉宾合影。

第一排：沈照理、Richard Lawrence Edwards、Ognjen Bonacci、林学钰、Derek Ford、袁道先、Petar Milanovic、薛禹群、汪集旸、姜玉池。

第二排：刘再华、徐永新、Andrzej Tyc 、Ralf Benischke、Chris Groves、潘根兴、蒋忠诚、王世杰、赵景波、唐常源、卢光辉、章程。

第三排：曹建华、Neven kresic、祁士华、Andrej Kranjc、Tran Tan Van、Francois Zwahlen、蔡运龙、余龙江、王建力。

IRCK's Daily Management Structure
国际岩溶研究中心日常管理结构

IRCK/IKG Director Jiang Yuchi hosted the IRCK Annual Work Meeting.

国际岩溶研究中心主任、岩溶地质研究所所长姜玉池研究员主持年度工作会议。

The director of IRCK appointed an executive deputy director and a secretary-general. Directors were appointed to the following divisions: secretariat (Zhang Cheng), financial (Zhong Kaiwei), international cooperation and training (Zhang Cheng), scientific research (Cao Jianhua), domestic affairs (Zhu Mingqiu), and consultation and outreach (Pei Jianguo).

国际岩溶研究中心主任在2009年9月任命了常务副主任和秘书长；同时设置中心秘书处（章程兼主任）、财务部（钟开威任主任）、国际合作交流与培训部（章程兼主任）、科研部（曹建华兼主任）、内联部（朱明秋任主任）、咨询与推广部（裴建国任主任）。

上左图 /Top left
IRCK Director Jiang Yuchi, director of the Institute of Karst Geology, professor of oceanic geology.
国际岩溶研究中心第一届主任姜玉池先生，岩溶地质研究所所长、海洋地质学家、教授。

上中图 /Top middle
IRCK Executive Deputy Director Cao Jianhua, professor of karst ecosystems.
国际岩溶研究中心常务副主任曹建华博士，岩溶生态系统教授。

上右图 /Top right
IRCK Secretary-General Zhang Cheng, professor of karst environment.
国际岩溶研究中心秘书长章程博士，岩溶环境学教授。

第二章　组织建设与管理

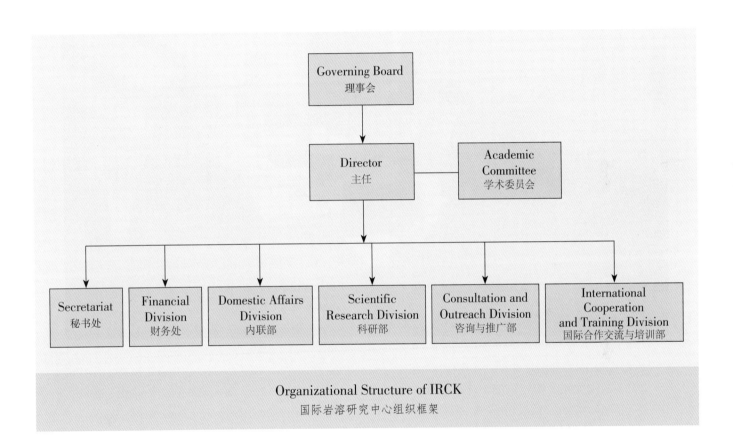

Organizational Structure of IRCK
国际岩溶研究中心组织框架

First Session of the 1st Governing Board
第一届理事会第一次会议

Before the first meeting of the 1st IRCK Governing Board, the chairman of the Governing Board, Dr. Wang Min, Vice Minister of China Ministry of Land and Resources convened a prelude in the foreign language reading room of IRCK's library. In attendance were representatives from UNESCO (Prof. Walter Erdelen), China Geological Survey (Prof. Zhong Ziran), Chinese Academy of Sciences (Prof. Dong Shuwen), IRCK (Prof. Jiang Yuchi), and others. They discussed IRCK's organizational structure, operation law, and the way to achieve IRCK's goals.

第一次理事会召开之前，国际岩溶研究中心理事会主席、国土资源部副部长汪民先生主持召开了预备会，会议在岩溶所图书馆外文阅览室举行，预备会就为实现国际岩溶研究中心目标、发挥其功能，以及国际岩溶研究中心的组织机构、运行规律模式等方面进行协商。

参加会议人员：联合国教科文组织代表 Walter Erdelen，国际岩溶研究中心理事会主席汪民，国际岩溶研究中心主管单位代表、国土资源部国际合作与科研司、中国地质调查局代表钟自然，中国地质科学院代表董树文及国际岩溶研究中心依托单位代表姜玉池。

第二章 组织建设与管理　033

The first session of the 1st IRCK Governing Board was held on December 13, 2008 in the guest room of the Karst Geology Museum of China. The meeting was hosted by IRCK Governing Board Chairman, Dr. Wang Min. During the meeting, Prof. Jiang Yuchi was appointed as IRCK Director, and Prof. Yuan Daoxian as director of the Academic Committee. IRCK's "Financial, Administrative and Personnel Management Procedures" were adopted.

2008年12月13日，国际岩溶研究中心第一届理事会第一次会议在岩溶所岩溶地质博物馆贵宾室召开，会议由理事会主席汪民先生主持，会议宣布了中心第一届理事会主席和理事成员名单，任命中国地质科学院岩溶地质研究所所长姜玉池为研究中心主任，任命中国科学院院士袁道先研究员为中心学术委员会主任，原则上通过《联合国教科文组织国际岩溶研究中心财务、行政和人员管理办法》。

Vice Minister Wang Min of China Ministry of Land and Resources met with Prof. Walter Erdelen, UNESCO Assistant Director-General for Natural Sciences on December 13, 2008, at the gate of IRCK's library.
Right to left: Walter Erdelen, Wang Min, Li Wei, Jiang Yuchi, Li Zhijian, Zhong Ziran.

2008年12月13日，国土资源部副部长汪民先生与联合国教科文组织助理总干事Walter Erdelen相遇岩溶所图书馆，并互致祝贺和感谢。
从右至左：Walter Erdelen、汪民、李薇、姜玉池、李志坚、钟自然。

国际岩溶研究中心 *6* 年历程

上左图
Top left

The Chairman of the IRCK Governing Board, Dr. Wang Min, hosted the meeting, and emphasized that the establishment of IRCK should be recognized as a landmark event for international karst science. He stated that China should give special consideration to the construction of IRCK, and that IRCK should strive for a good working plan during the initial phase of its establishment. He also emphasized the importance of establishing good relationships with the Institute of Karst Geology, key laboratory of Karst Dynamics and other organizations both within China and abroad.

国际岩溶研究中心理事会主席主持会议,并强调中心要充分认识中心成立的重要性及其意义,以更加开放的姿态,梳理工作思路,处理好中心、岩溶所、重点实验室的关系,开展国内外合作交流与多学科交叉研究、开展咨询与培训。

上右图
Top right

Dr. Robert Missotten, representing member of the IRCK Governing Board from UNESCO, spoke at the meeting. He thanked the Chinese Government for their great efforts for IRCK. He emphasized the objectives and functions of IRCK, and IRCK should promote and spread the karst dynamic system to understand the kinds of global karst dynamic system, and provide scientific and technical supports for sustainable development of the fragile ecosystem in karst areas. He also recommended that IRCK should set up the academic committee so as to strengthen ties with UNESCO, improve international cooperation and reflect internationalism, particularly strengthening cooperation with African countries.

罗伯特·米索腾是代表联合国教科文组织的中心理事会委员,他在会上的发言,首先,感谢中国政府对岩溶中心启动工作所付出的巨大努力;其次,强调了中心的目标和职能,在全球范围内推广岩溶动力系统,了解全球岩溶动力系统类型,为岩溶地区脆弱的生态系统的可持续发展提供科学与技术支撑;最后,提出中心要尽快设立中心学术委员会,加强与联合国教科文组织联系、加强国际合作,体现中心国际性,加强与非洲国家代表之间的合作。

The IRCK Governing Board Chairman presented certificates to the 13 members of the Governing Board.
Top left: Prof. Chris Groves received a certificate.
Top right: Prof. Yuan Daoxian received a certificate.
Bottom left: Dr. Robert Missotten received a certificate.
Bottom right: Prof. Wang Yanxin received a certificate.

国际岩溶研究中心理事会主席为到会的13位委员颁发聘书。
上左：为克里斯·格若斯委员颁发聘书；
上右：为袁道先委员颁发聘书；
下左：为罗伯特·米索腾委员颁发聘书；
下右：为王焰新委员颁发聘书。

Second Session of the 1st Governing Board
第一届理事会第二次会议

Mr. Jiang Jianjun, Director of the Department of International Cooperation and Science and Technology, MLR, hosted the meeting, on behalf of IRCK Governing Board Chairman Wang Min.

国土资源部科技与国际合作司司长姜建军先生，代表国际岩溶研究中心理事会主席汪民先生到会并主持会议。

The second session of the 1st Governing Board was held on December 5, 2009 in the guest room of the Karst Geology Museum of China at Institute of Karst Geology(IKG). Thirteen board members took part in the meeting, and three main subjects were under discussion:

1. Dr. Robert Missotten presented certificates to four board members: Dr. Wilhelm Struckmeier, Prof. Derek C. Ford, Prof. Petar T. Milanovic, Prof. Jiang Jianjun.

2. IRCK's Annual Report for 2009 and Work Plan for 2010 were presented and reviewed.

3. The IRCK Development Plan (2010–2015) was presented and reviewed.

Finally, after discussion it was decided that meetings of the Governing Board will be held once every two years.

第二章 组织建设与管理

From left to right: Jiang Yuchi, Dong Shuwen, Lu Yaoru, Derek C. Ford, Yuan Daoxian, Petar T. Milanovic, Jiang Jianjun, Robert Missotten, Zhang Hongren, Wilhelm Struckmeier, Wang Jiyang, Chris Groves, Huang Junhua, Jiang Shijin.

2009年12月5日，第一届理事会第二次会议在岩溶地质研究所中国岩溶地质馆贵宾厅召开，会议主要有3个议题：(1) 罗伯特·米索腾博士代表理事会为4位理事会成员（威赫穆·斯图克梅尔，德瑞克·福特，彼塔·米兰诺维奇，姜建军）补发聘书；(2) 听取和评议了《联合国教科文组织国际岩溶研究中心2009年的主要工作及2010年工作计划》；(3) 听取和评议了《联合国教科文组织国际岩溶研究中心发展计划（2010–2015年）》。最后经过提议和讨论，原则通过理事会、学术委员每2年召开一次。参加第二次会议的委员和代表共13人。

由左到右：姜玉池、董树文、卢耀如、德瑞克·福特、袁道先、彼塔·米兰诺维奇、姜建军、罗伯特·米索腾、张宏仁、威赫穆·斯图克梅尔、汪集旸、克里斯·格若斯、黄俊华、蒋仕金。

In the meeting, many board members emphazied on IRCK's actions towards developing scientific research and guiding karst scientists to develop the karst dynamic system worldwide. They largely agreed to simultaneous development of basic research and applied research on the karst dynamic system, including operation law, types and distribution of the karst dynamic system. The applied research should include large water resources dams and conservancy projects and integrating control for rocky desertification, fighting drought and flood, karst collapse, etc.

此次理事会，委员更多地关注了国际岩溶研究中心就如何开展科学研究，影响和引领各国岩溶科学家开展岩溶动力系统研究，比较一致的意见是岩溶动力系统的基础研究与应用研究要同时开展，包括系统的运行规律和类型及分布，应用性研究包括石漠化综合治理、旱涝灾害、大型水利工程、岩溶塌陷等。

Dr. Robert Missotten gave an important speech, expressing the opinion that since its establishment IRCK has put forth great effort and achieved a good beginning for the five objectives of IRCK. He especially congratulated IRCK for the success of the International Training Course on Karst Hydrogeology and Karst Ecosystem, and admired the Man and Biosphere special edition on Karst Rocky Desertification guest-edited by IRCK.

罗伯特·米索腾博士在会上做重要讲话，认为中心成立一年来，针对国际岩溶研究中心5个目标，做出了较大的努力，取得了较好开端，特别对"岩溶水文地质与生态"国际培训班的成功举办表示祝贺，对中心与中国人与生物圈国家委员会主办的《人与生物圈》编辑部合作出版的"岩溶石漠化"专辑特刊表示赞赏。

第二章　组织建设与管理　039

上左图 / Top left
Robert Missotten, representing the Governing Board, presented a certificate to Prof. Derek C. Ford.
罗伯特·米索腾博士代表理事会向委员德瑞克·福特颁发证书。

上右图 / Top right
Prof. Wang Jiyang gave suggestion on IRCK's progress in developing research on the karst dynamic system.
汪集旸委员就国际岩溶研究中心如何开展岩溶动力系统研究提出建议。

Board members listened carefully to a report on IRCK's annual work and the work plan for the next year. Right to left: Wilhelm Struckmeier, Yuan Daoxian, Derek C. Ford, Petar T. Milanovic, Cao Jianhua.

委员们认真听取中心的年度工作报告以及下年的工作计划。
从右到左：威赫穆·斯图克梅尔、袁道先、德瑞克·福特、彼塔·米兰诺维奇、曹建华。

国际岩溶研究中心 6 年历程

Third Session of the 1st Governing Board
第一届理事会第三次会议

Executive deputy director of the Chinese Academy of Geological Sciences, Wang Xiaolie hosted the meeting on behalf of IRCK Governing Board Chairman Wang Min.

中国地质科学院常务副院长王小烈先生，代表国际岩溶研究中心理事会主席汪民先生到会并主持会议。

The third session of the 1st Governing Board was held in Guilin on December 5, 2011. Ten board members of the Governing Board were in attendance: Derek Ford, Chris Groves, Petar Milanovic, Robert Missotten, Yuan Daoxian, Zhang Hongren, Wang Jiyang, Lu Yaoru, Lian Changyun (representing Zhong Ziran), Sun Baoliang (representing Jiang jianjun).

2011年12月5日上午，联合国教科文组织国际岩溶研究中心一届理事会三次会议在桂林顺利召开，10位理事出席会议，他们是Derek Ford、Chris Groves、Petar Milanovic、Robert Missotten、袁道先、张宏仁、汪集暘、卢耀如、连长云（代表钟自然）、孙宝亮（代表姜建军）。

Dr. Robert Missotten, representing UNESCO, gave an address at the meeting. He stated his admiration for the Category II center annual meeting system convened by the National Commission of the PRC for UNESCO. He expressed his interest in the construction and development of IRCK, and fully affirmed the work and achievements of IRCK during the past two years. He shared his hopes that IRCK would expand outreach and communication efforts, strengthen the karst carbon sink monitoring and international cooperation, continue to strengthen cooperation and contact with the IHP and IGCP, develop trans-boundary aquifer management projects and participate in IGCP's 40th anniversary activities and the 34th International Geological Congress.

罗伯特·米索腾博士首先代表联合国教科文组织致辞，他赞赏 UNESCO 中国国家委员会召开的二类中心年度例会制度，也非常关心国际岩溶研究中心基地的建设和发展，充分肯定了中心成立两年来的工作与成绩，希望岩溶研究中心扩大宣传力度，加强岩溶碳汇监测国际合作，继续加强与 IHP、IGCP 的合作与联系，开展跨边界含水层管理项目，参与 IGCP 成立 40 周年活动及 34 届国际地质大会。

IRCK Director Jiang Yuchi introduced IRCK's accomplishments, IRCK Executive Deputy Director Cao Jianhua reported on IRCK's activities in the past two years (2010–2011) and IRCK Working Plan (2012). IRCK Secretary-General Zhang Cheng reported on the IRCK Development Plan (2010–2015). The Board members reviewed IRCK's 2010–2011 activities, IRCK Working Plan (2012), and the IRCK Development Plan (2010–2015). They also gave advice on future construction and development. It was agreed that the Governing Board and Academic Committee sessions should generally be held on the first weekend in December every odd year. The Governing Board advised IRCK to pay more attention to international cooperation for trans-boundary aquifer management, promote the establishment of the global karst process and carbon sink monitoring network, and further accelerate the capacity-building to the IRCK powerful team.

国际岩溶研究中心主任姜玉池研究员介绍中心工作，常务副主任曹建华博士报告中心 2010 – 2011 年工作进展与 2012 年工作计划，秘书长章程博士报告了中心 2010 – 2015 年发展规划。随后委员们对中心 2010 – 2011 年工作进展与 2012 年工作计划、2010 – 2015 年中心发展规划进行了评议，提出未来建设和发展的建议。原则上确定理事会和学术委员会会议固定在公历逢单年度 12 月的第一个周末召开。国际岩溶研究中心理事会建议中心要重视跨边界含水层管理国际合作的工作；做好全球岩溶碳汇监测网建立的宣传和建站的工作；进一步加快中心以人才队伍建设为中心的能力建设。

The First Session of the 1st IRCK Academic Committee
第一届学术委员会第一次会议

The first session of the 1st IRCK Academic Committee was held at the Karst Geology Museum of China on December 13, 2009. It was attended by 27 committee members. At the meeting, the director of Academic Committee was appointed, and two Vice directors were elected. The Chairman also explained the production of the name list of the Academic Committee members. Both the Academic Committee Constitution and the IRCK Development Plan (2010–2015) were examined and adopted. Finally, the IRCK Director and the director of the IRCK Academic Committee presented certificates to the Academic Committee members.

2009年12月13日，IRCK第一届学术委员会第一次会议在中国岩溶地质馆贵宾厅召开，27名委员和相关人员参加了会议。会议主要有3个议程：(1) 宣布中心第一届学术委员会主任、选取2名副主任，学术委员会主任对学术委员会委员名单的产生做说明；(2) 审议并通过学术委员会的章程、审议中心的发展规划（2010 – 2015年），提出修改建议；(3) 中心主任、中心学术委员会主任共同颁发委员聘书。

左六图 / Left six

IRCK Director Jiang Yuchi and IRCK Academic Committee Chairman Yuan Daoxian presented certificates to committee members.

中心主任姜玉池教授、学术委员会主任袁道先院士为委员们发聘书。

右六图 / Right six

During the session, IRCK Academic Committee members read reports, gave advice and comments.

在学术委员会上委员们认真阅读文件、踊跃发言、献计献策。

IRCK was established in Guilin, China. It is not only due to the regional advantages of China karst research, the efforts of Chinese karst researcher, and the Chinese Government's commitment to the problems of karst resources and environment and relevant support; but also a good platform for the international karst academic community, where academic thinking and progress, major issues in consultation, cooperative projects, and so on, can unfold.

The members of the IRCK Academic Committee are renowned international karst experts and representatives from karst organization around the world, who have taken an active part in IGCP programs for many years. The IRCK Academic Committee brought these experts together in the international karst capital – Guilin, to discuss the construction and deployment of the first geoscience center under the framework of IGCP. The atmosphere of the meeting was very positive and friendly.

国际岩溶研究中心在中国桂林成立，不仅与中国岩溶研究的地域优势、中国岩溶研究者的努力、中国政府对岩溶资源、环境问题的关注和相关支持密切相关，更重要的是该中心的成立，给国际岩溶学术界提供了一个很好的平台，在这个平台上可以更充分地开展学术思维和进展、重大问题咨询、项目合作等。

学术委员会委员主要来自长期、积极参加 IGCP 项目，不同国家岩溶组织代表、国际著名的岩溶学家，通过学术委员会，聚集在国际岩溶之都——桂林，一起讨论 IGCP 框架下的第一个地学中心的建设部署、发展规划，心情非常愉快，从会前、会间到会后，大家都沉浸在愉悦、和谐的气氛中。

The Second Session of the 1st IRCK Academic Committee
第一届学术委员会第二次会议

The meeting was hosted by Prof. Yuan Daoxian,
IRCK Academic Committee Director and CAS academician.
国际岩溶研究中心学术委员会主任、中国科学院院士袁道先教授主持会议。

The second session of the 1st Academic Committee of IRCK was held at the Institute of Karst Geology in Guilin on December 3 – 4, 2011. Eighteen committee members attended the meeting, including IRCK's director, executive deputy director, and secretary-general. The meeting focused on the annual work and the work plan for 2012, establishing a global karst process and carbon sink monitoring network, and the Development Plan (2010–2015).

The Academic Committee gave advice to IRCK: (1) deepening understanding of the karst process and carbon sink, and data collection; (2) enhancing the process and response of karst water resources and water cycle to global climate change; (3) actively joining in and developing the trans-boundary aquifer management.

2011年12月3-4日，国际岩溶研究中心学术委员会第一届第二次会议在中国地质科学院岩溶地质研究所召开。中心学术委员会18位成员参加会议，中心主任、常务副主任和秘书长列席会议。会议主要内容是中心年度工作及下年度工作计划，关于建立全球碳汇监测网的提议与邀请，中心"2010-2015年发展规划"等征求委员意见与建议。

委员们一致建议，加强对岩溶碳汇过程的理解和数据的积累；在全球气候变化的大背景下，加强岩溶水资源、水循环过程与响应研究，积极参与和开展跨边界含水层管理研究。

During the session, two academic sessions were hosted by Prof. Wang Jiyang and Prof. Jiang Zhongcheng, December 3–4, 2011. Prof. Lu Guanghui, Prof. Ognjen Bonacci, Prof. Ralf Benischke, Prof. Yongxin Xu, Dr. Andrzej Tyc, Prof. Chris Groves, Prof. Liu Zaihua and Dr. Tran Tan Van gave presentations at the meeting. The major topics included: the effective utilization of rainwater flooding in lowlands, groundwater recharge estimation, deep caves in Poland, travertine degradation in Huanglong and the importance of monitoring.

3日下午和4日下午，学术委员会安排了两场学术报告会，分别由汪集旸院士和蒋忠诚研究员主持，中心学术委员会八位委员：卢光辉（Andrew Lo）、Ognjen Bonacci、Ralf Benischke、Xu Yongxin、Andrzej Tyc、Chirs Groves、Liu Zaihua、Tran Tan Van 做了学术报告。内容主要涉及雨水的有效利用与低地洪涝防治、地下水的补给估算、深部洞穴（波兰）、黄龙钙华退化成因及表层带监测作用与意义等。

IRCK Annual Work Meeting
国际岩溶研究中心年度工作会议

  The annual meeting concentrated on reports from IRCK's budget and organizational management, scientific research, academic exchange, consulting service and science dissemination, training course and improvements to operational conditions during the last year. The work plan and budget for the next year were also discussed. The meeting was attended by IRCK/IKG Director Jiang Yuchi, IRCK Academic Committee director Yuan Daoxian, IRCK Executive Deputy Director Cao Jianhua, IRCK Secretary-General Zhang Cheng, and the heads of each IRCK and IGK division.

  The IRCK annual work meeting was usually held at the first week in March every year.

  年度工作会议主要内容包括：上一年度中心在组织管理、科学研究、学术交流、科普与咨询、国际培训及运行条件完善等方面的工作报告、经费使用；下一年度主要工作安排及经费预算。与会中心成员评议，并提出完善、修改意见。

  参会人员主要包括中心主任、学术委员会主任、常务副主任、秘书长、中心各分工部门的负责人、秘书处人员，同时邀请岩溶地质研究所相关领导和责任人参加。

  国际岩溶研究中心的年度工作会议在每年3月的第一个星期召开。

Chapter 2 Organization and Management

IRCK's "Financial, Administrative and Personnel Management Procedures" were passed by the first session of the 1st Governing Board. The "Constitution of the Academic Committee of IRCK" was passed by the first session of the 1st Academic Committee. IRCK's "statutes" were examined and amended by IGCP. The "IRCK Development Plan (2010–2015)" was considered by the Academic Committee, and passed by the Governing Board in 2011.

The International Research Centre on Karst
Under the Auspices of United Nations Educational,
Scientific and Cultural Organization (UNESCO)

Statutes
(Draft)

Article 1 Participation

1. The International Research Centre on Karst (hereinafter referred to as "the IRCK") shall be an institution under the auspices of United Nations Educational, Scientific and Cultural Organization (hereinafter referred to as UNESCO) (category 2), established as one of the international institutes of the Chinese Academy of Geological Sciences (CAGS) and in accordance with "the Agreement between the Government of the People's Republic of China and the United Nations Educational, Scientific and Cultural Organization (UNESCO) Concerning the Establishment and Operation of the International research center on karst in Guilin, China, under the Auspices of UNESCO" (hereinafter referred to as "the Agreement"), at the service of Member States of UNESCO, which, by their common interest in the objectives of the IRCK and their commitment to improved environmental and resources management in karst region, desire to cooperate with the IRCK.

2. Member States of UNESCO wishing to participate in the IRCK's activities, as provided for under these Statutes, shall send the Director-General of UNESCO a notification to this effect. The Director-General shall inform the IRCK and the notifying Member State(s) of the receipt of such notification.

Annex V:

The International Research Centre on Karst
Under the Auspices of UNESCO

Financial, Administrative and Personnel Management Procedures

Chapter I General Provisions

Article 1

The International Research Centre on Karst under the auspices of UNESCO (hereinafter referred to as "the IRCK") is established in accordance with *"the Agreement between the Government of the People's Republic of China and the United Nations Educational, Scientific and Cultural Organization (UNESCO) concerning the establishment and operation of the International Research Center on karst in Guilin, China, under the auspices of UNESCO"*.

The IRCK shall act under the Chinese law as one of the international institutes of the Chinese Academy of Geological Sciences.

In order to standardize and strengthen the building and operational administration of the Center, the following management procedures are formulated.

Article 2

The IRCK shall be at the service of Member States and Associate Members of UNESCO. The main Objectives and functions of IRCK are to organize and implement international scientific and technological cooperation in karst, provide a platform for the exchange of scientific information about karst dynamics, the sustainable utilization of karst resources and eco-environmental protection; develop a network of karst experiment and

12

由教科文组织赞助的
国际岩溶研究中心（IRCK）
章程（草案）

第1条　参与工作

1. 国际岩溶研究中心（以下简称"中心"）为一个独立自主的由教科文组织赞助的第2类机构，它根据中国国土资源部与联合国教科文组织签署的《中华人民共和国政府与联合国教育、科学及文化组织关于在中国桂林建立由教科文组织赞助的国际岩溶研究中心及其运作的协定》（以下简称《协议》）而建立，为那些对本中心的目标有共同兴趣，为了解决岩溶地区的环境和资源问题希望与其开展合作的教科文组织会员国提供服务。

2. 希望按《章程》之规定参与本中心活动的教科文组织会员国，须向教科文组织总干事呈交一份表明此意愿的通知。总干事把该通知知悉一事通知本中心以及有关会员国。

第2条　法人资格

3. 根据《协议》，IRCK中心在中国享有行使其职能所需的法人资格和法定能力，尤其享有：
　a）订立合同、进行诉讼、购置和转让动产和不动产的法定能力；
　b）接受补助金、收取劳务费和获取一切必要手段所需的法定能力。

第3条　目标与职能

4. IRCK中心的目标是：
　(a) 通过科学研究、出版活动和国际合作，促进岩溶动力学的发展；
　(b) 促进国际合作与交往，为世界各地属于教科文组织国际地球科学计划、国际地质学联盟（IUGS）岩溶委员会以及国际水文

附件五

**由联合国教科文组织赞助的国际岩溶研究中心
财务、行政和人员管理办法
（草案）**

第一章　总　则

第一条　国际岩溶研究中心（以下简称中心）依据《中华人民共和国政府与联合国教育、科学及文化组织关于在中国桂林建立由教科文组织赞助的国际岩溶研究中心及其运作的协定》设立。中心作为中国地质科学院的国际机构之一，依据中国法律开展活动。为规范和加强中心的建设和运行管理，制定本办法。

第二条　中心向合作的教科文组织会员国提供服务。主要目标和任务是组织实施国际岩溶科技合作活动，搭建有关岩溶动力学、岩溶资源可持续利用和生态环境保护的科学信息交流平台，建立岩溶实验和示范基地网、提供咨询服务、技术信息和人员培训，传播岩溶科学知识和为制定石漠化治理和生态恢复综合方案奠定基础，以保持脆弱岩溶环境的良性生态循环，促进岩溶地区可持续发展。

第二章　机构、人员与职责

第三条　中心的组织框架包括学术委员会、秘书处、财务部、国际合作交流与培训部、内联部、咨询与推广部、科研部等。

第四条　中心主任

　1、中心主任由中国地质科学院上级主管领导与教科文组织总干事和理事会协商后任命，原则上由中国地质科学院岩溶地质研究所所长担任，并主持中心的工作。

　2、中心主任履行以下职责：
　(1) 按照理事会审定的计划和指示管理中心；

为了国际岩溶研究中心的正常运行，中心先后编制了《联合国教科文组织国际岩溶研究中心财务、行政和人员管理办法》（第一届理事会第一次会议审议通过），《学术委员会章程》（第一届学术委员会第一次会议审议通过），《国际岩溶研究中心章程》（2009年5月经联合国教科文组织IGCP审查、修改），《国际岩溶研究中心发展规划》（2010–2015）(2011年学术委员会审议、理事会通过)。

IRCK Regular Publication
国际岩溶研究中心日常出版物

IRCK produced quarterly and annual reports. These were some of the most important publications during the first six years of IRCK. The quarterly newsletters were usually completed in the last month of each quarterly season, and the annual reports were published in February every year.

Quarterly newsletters and annual reports PDF versions were submitted to UNESCO, CAGS, CGS, MLR and Board members and committee members.

国际岩溶研究中心季度报、年报是中心在第一个6年建设过程中最为重要的出版物之一，通常季度报在上季度后1个月内编写完成，年报则在每年的2月完成，并印刷。

季度报、年报完成后，电子版(PDF)通过邮件发送中心主管部门及理事会、学术委员会委员。

第二章　组织建设与管理

IRCK Website
国际岩溶研究中心网页制作和运行

In order to improve both outreach and contact members, IRCK website was produced. The website facilitates organizing international symposia and conferences, develops the annual international training courses, and publicizes news and academic progress.

为了更好地宣传和联系，国际岩溶研究中心制作了中心网站，利用网站发表新闻、组织国际会议、开展年度国际培训班、报道最近学术研究进展等，为中心的运行带来较大的便利。

International Joint Research Center on Karst Dynamic System and Global Change approved as a National Center for International Research by Chinese Government
"岩溶动力系统与全球变化国际联合研究中心"被中国政府批准为国家级国际联合研究中心

The International Joint Research Center on Karst Dynamic System and Global Change (IRCKDSGC) was designated by the Ministry of Science and Technology (MOST) of the People's Republic of China as a National Center for International Research in September 2013. IRCKDSGC, hosted by IKG and IRCK, was selected and recommended by the Department of Science and Technology of Guangxi Zhuang Autonomous Region, and approved and designated by MOST.

Since the inception of IRCK, the research team led by Prof. Yuan Daoxian, CAS academician has made significant contributions to international karst research, and has won recognition from the international karst academic community. Specifically, IRCK researchers have undertaken extensive international exchange, training programs and long-term international cooperation, and have achieved innovative results in terms of the karst dynamic system and global change. They have also established a basis for international karst comparative studies, trained young researchers in cooperation with foreign universities, and recommended young scientists to work in international karst academic organizations and raise their profiles in international academia.

The objectives of IRCKDSGC are to:

1) share the world's latest developments in karst dynamic system and global change research through international science and technology cooperation;

2) leverage international science and technology resources to serve the national social and economic development;

3) achieve innovative outcomes in four areas
   – the significance and carbon sink effects of the karst dynamic system and karst process in the global carbon cycle;
   – high-resolution paleoclimate records from stalagmites;
   – effective management of karst aquifers in extreme weather conditions;
   – the response of fragile karst ecosystems to global change;

4) take the lead and demonstrate how to address the resource and environmental issues in karst areas in developing countries.

2013年9月，科学技术部认定"岩溶动力系统与全球变化国际联合研究中心"为国家级国际联合研究中心。

岩溶动力系统与全球变化国际联合研究中心是以中国地质科学院岩溶地质研究所/联合国教科文组织国际岩溶研究中心为依托，经广西壮族自治区科学技术厅遴选、推荐，国家科技部审批认定的。

国际岩溶研究中心自成立以来，以袁道先院士为首席科学家的岩溶研究团队对国际岩溶研究做出了重要贡献，得到了国际岩溶学术界的认可；开展了广泛的国际交流、培训与长期的国际合作，在岩溶动力系统与全球变化方面取得了创新性成果；建立了国际岩溶对比研究基地；与国外大学联合培养年轻人；推荐年轻学者，在国际岩溶学术组织任职，提高他们在国际学术界知名度。

岩溶动力系统与全球变化国际联合研究中心通过国际科技合作，分享当今世界最新的资讯和成果；积极利用国际科技资源，服务国家经济社会发展的大局；同时在岩溶动力系统运行规律与岩溶作用对全球碳循环的意义及碳汇效应、石笋高分辨率古气候记录、应对极端气候岩溶含水层管理和脆弱岩溶生态系统对全球变化的响应等四个方面取得创新性成果，尤其是在发展中国家的岩溶区资源环境问题对策，起引领和示范作用。

Prof. Derek C. Ford, Fellow of the Royal Society of Canada; Mr. Han Qunli, director of the Division of Ecological and Earth Sciences, UNESCO; Mr. Wang Yan, deputy director general of CGS; and Prof. Yuan Daoxian, CAS academician (from left to right) jointly unveiled the nameplate for International Joint Research Center on Karst Dynamic System and Global Change.

中国地调局常务副局长王研教授、联合国教科文组织地学生态部主任韩群力博士、加拿大科学院院士及国际洞穴协会前主席 Derek C. Ford 教授、中国科学院院士袁道先教授为国家级国际联合研究中心揭牌。

Photo caption:
Name: Wulong Geopark, world natural heritage site
Location: Chongqing
Time listed in Geopark: 2004
Time listed in World Heritage: 2007
Summary: Wulong Geopark, geological heritage site is featured by carbonate karst landscape, geomorphologic features such as caves, sinkholes group, natural bridge group, sinkhole, canyons, bare rock, fengcong, fenglin, underground stream, hot springs, it is a rare karst wonders in China.

照片说明：
名称：武隆地质公园、世界自然遗产地
所在地点：重庆
列入地质公园时间：2004年
列入世界遗产时间：2007年
概述：武隆地质公园的地质遗迹和地质景观以碳酸盐岩溶地貌为特色，其溶洞群、天坑群、天生桥群、竖井群、峡谷、地缝、石林、石芽、峰丛、峰林、地下伏流、间歇泉、温泉等各类地貌分布广泛，组合完好，种类齐全，在全国目前发现的喀斯特地貌奇观中实属罕见。

Chapter 3

Scientific Research

第三章　科学研究

At the opening ceremony on establishing International Research Center on Karst (IRCK), UNESCO Assistant Director-General proposed that karst dynamic system should be promoted and applied globally to provide scientific and technological support for the economic and social development in karst regions. Therefore, IRCK mainly focuses its scientific research on karst dynamic system. Given the integrity of the study on karst dynamic system, this chapter is mainly composed of the research findings since the inception of IRCK (2008-2013), as well as some results achieved in the process of implementing karst-related IGCP projects. They are mainly as the following areas:

(1) the concept of karst dynamic system;

(2) the characteristics of karst dynamic system ;

(3) the research contents and its application of karst dynamic system ;

(4) the research projects and their major progress:

Carbon cycle: karst dynamic system and global climate change;

Water cycle: karst dynamic system and water cycle and water resources;

Calcium cycle: the foundation for karst ecosystem research;

Karst dynamic system and sustainable development;

(5) international cooperation projects;

(6) overview of the field research demonstration.

在国际岩溶研究中心成立大会上，联合国教科文组织助理总干事提出，要在全球范围推广和应用岩溶动力系统，为岩溶区的经济社会发展提供科技支撑。因此，国际岩溶研究中心的科学研究主要围绕岩溶动力系统展开。考虑到岩溶动力系统研究的完整性，本章节的内容主体为国际岩溶研究中心成立以来（2008-2013）的研究成果，结合部分与岩溶相关IGCP项目执行过程中取得的成果，主要包括以下几个方面：

(1) 岩溶动力系统的概念；

(2) 岩溶动力系统的特点；

(3) 岩溶动力系统研究内容与应用；

(4) 科研项目与主要进展：

  碳循环：岩溶动力系统与全球气候变化；

  水循环：岩溶动力系统与水循环、水资源；

  钙循环：岩溶生态系统研究的基础；

  岩溶动力系统与可持续发展；

(5) 国际合作项目；

(6) 野外研究基地建设介绍。

Karst Dynamic System: Advanced Idea in Karst Research
岩溶动力系统：现代岩溶研究的新思路

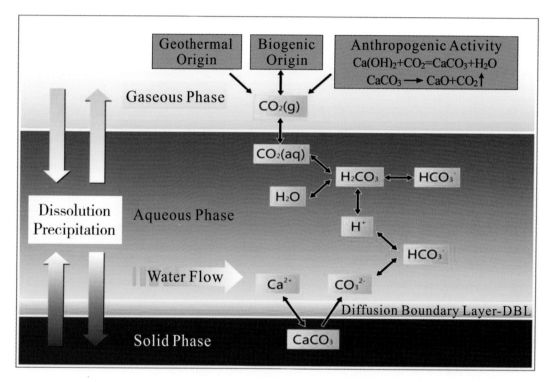

Structure chart of karst dynamic system
岩溶动力系统结构图

  A karst dynamic system involves the transfer of energy and matter within the carbon, water and calcium (and magnesium) cycle. It occurs at the interfaces of the lithosphere, hydrosphere, atmosphere and biosphere and controls the formation and evolution of karst, but is moderated by the exsiting formed karst features.

  岩溶动力系统可定义为控制岩溶形成演化，并常受制于已有岩溶形态的，在岩石圈、水圈、大气圈、生物圈界面上的，以碳、水、钙循环为主的物质、能量传输、转换系统。

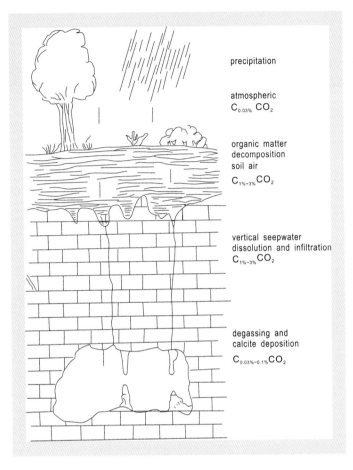

左图
Left

Diagram of carbon migration in karst dynamic system
岩溶动力系统碳迁移示意图

岩溶动力系统是一个开放的三相开放不平衡系统，与地球的"四圈层"密切联系，其基本特征是对环境反应敏感。通过前期的研究，我们已掌握岩溶动力系统有四大功能：

（1）驱动各种岩溶形态的产生，并通过其所造成的地表地下双层岩溶空间结构和碱性地球化学背景导致一系列环境问题，如旱、涝、石漠化、地面塌陷、生物多样性受限等；

（2）通过岩溶作用由大气回收或向大气释放 CO_2，调节大气温室气体浓度，缓解环境酸化；

（3）驱动元素迁移、富集、沉淀，形成有用矿产资源，影响生命；

（4）记录全球环境变化过程，由于岩溶动力系统与全球四圈层的密切关系，它可以敏感地反映并真实地记录各种环境因子，包括降雨量、气温、植被等变化，为研究全球变化提供依据。

Karst dynamic system is a triphase open disequilibrium system characterized by a high sensitivity to climate and environmental change. The basic functions of the karst dynamic system are:

(1) to drive the formation of karst features and related environmental problems, such as drought, water-logging, rocky desertification, subsidence and biodiversity constraint;

(2) to contribute to the regulation of the greenhouse gases in the atmosphere and to the mitigation of environmental acidification;

(3) to drive the migration, enrichment and precipitation of certain elements and thus influence both the formation of mineral deposits and the biodiversity of life in karst areas;

(4) to record the course of environmental change. As karst dynamic system is closely associated with Four Great Earth Realms, it can sensitively respond to and faithfully record various environmental factors, including rainfall, air temperature and vegetation changes, so as to provide basis for the study on global change.

Karst Ecosystem: the Development of Karst Dynamic System
岩溶生态系统：岩溶动力系统的发展

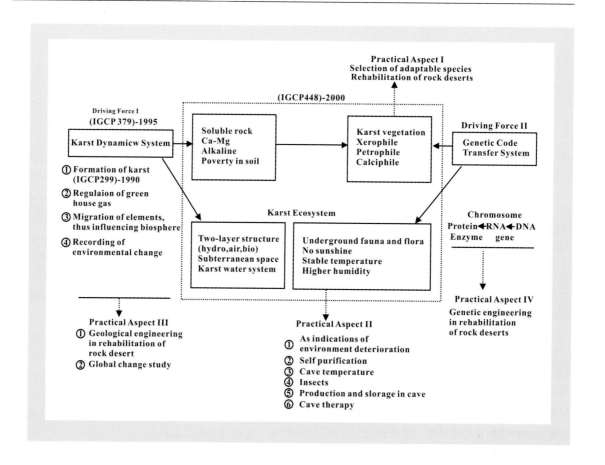

The structure, driving mechanism and function of the karst ecosystem
岩溶生态系统的结构、驱动机制和功能框图

The karst ecosystem can be explained as the ecosystem that is restrained by the karst environment. Its connotation includes how karst environment affects life and also the reaction of life to karst environment.

The operation of a karst ecosystem is driven by two systems: the Karst Dynamic System (KDS) in the abiotic aspect, and the Genetic Code Transfer System (GCTS) in the biotic aspect.

The Karst Dynamic System involves the transfer of matter and energy within the carbon, water and calcium (and magnesium) cycles. It occurs at the interfaces of the lithosphere, hydrosphere, atmosphere and biosphere, controls the formation of two features for inorganic environment within karst ecosystem (i.e., carbonate rocks that can be easily dissolved, and calcium-rich, alkaline and infertile soils; two-layer hydro-geological structure and underground space), and influences life.

The key function of the Genetic Code Transfer System is to transfer various genetic information in karst areas from DNA through RNA to protein. It controls the formation and evolution of the community of special producers, consumers and decomposers living in surface habitats that have no soil lacking of water but rich in calcium and underground habitats that have no light but humid and constant temperature. They combine to form karst ecosystem.

岩溶生态系统是受岩溶环境制约的生态系统，是岩溶地质与生态学的交叉渗透，是岩溶动力学的发展。

岩溶生态系统的运行受到两个系统联合作用的驱动，即无机方面的"岩溶动力系统"和生命方面的"遗传信息传递系统"。

岩溶动力系统是在大气圈、水圈、岩石圈和生物圈，以碳、水、钙（及其他金属元素）循环为主要形式的物质能量传输系统，并受到已有岩溶形态的影响。它制约岩溶生态系统中的无机环境方面两个特点（即易溶的碳酸盐岩和富钙、偏碱性、贫瘠的土壤，以及双层水文地质结构和地下空间）的形成，并影响生命。

遗传信息传递系统则通过遗传信息，按照中心法则由DNA到RNA到蛋白质的传递，制约着无土、缺水、富钙的地面环境和无光、潮湿、相对恒温的地下环境的特殊生产者、消费者和分解者群落的形成和演化，从而构成岩溶生态系统。

Characteristics of Karst Dynamic System
岩溶动力系统特点

Sensitivity of Karst Dynamic System to Climate Change
岩溶动力系统对气候环境变化响应的敏感性

The automatic monitoring instruments revealed that water level, water temperature, pH, HCO_3^- concentration and partial pressure of CO_2 of groundwater. Maocun karst underground river responded sensitively to the three storm events in late June, 2010. This reflects that karst process and carbon cycle also responds sensitively to rainfall events.

Three storms: 33 mm for storm 1 (06:30–11:30 on 13 June), 62.5 mm for storm 2 (03:30–13:00 on 14 June), 88.5 mm for storm 3 (08:45–17:15 on 16 June).

通过自动化监测仪器，揭示桂林毛村岩溶地下河中对反映岩溶作用及碳迁移的指标 pH、HCO_3^- 浓度及通量和水体中的 CO_2 分压对 2010 年 6 月中下旬 3 场降雨事件响应敏感。

3 场降雨事件：storm 1 的降雨量为 33 mm（6 月 13 日 06:30 – 11:30），storm 2 的降雨量为 62.5 mm（6 月 14 日 03:30 – 13:00），storm 3 的降雨量为 88.5 mm（6 月 16 日 08:45 – 17:15）。

对页图
Opposite page
Sensitivity of karst dynamic system to climate change
岩溶动力系统对气候环境变化响应的敏感性

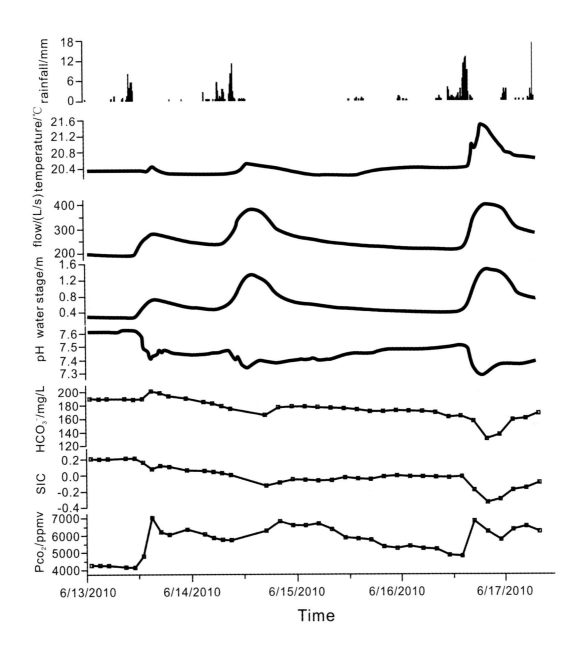

Calcium-Richness and Alkalinity in Karst Dynamic System
岩溶动力系统的富钙、偏碱性

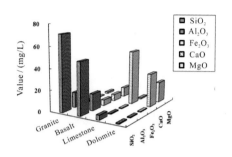

碳酸盐岩的富钙性 Carbonate rock-richness in soil

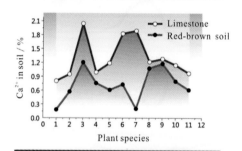

土壤富钙性 Calcium-richness in soil

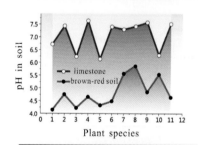

土壤的偏碱性 Alkalinity in soil

Carbonate rocks are the material basis of Karst Dynamic System (KDS), therefore, carbonate rocks with rich calcium result in the calcium-rich in karst dynamic system. This can be obviously found in the atmosphere, hydrosphere and biosphere in karst areas: the calcium content of rainwater in Guangxi and Guizhou karst regions is recorded to be 2.9–6 mg/L, while that in non-karst regions is often less than 1 mg/L; the calcium ion content in karst water, especially in karst groundwater, reaches 50–120 mg/L, while that in exogenous water of non-carbonate rock areas contains only from almost null to a dozen mg/L; the vegetation in karst areas contains higher lime, calcium and magnesium content, while its aluminium content is lower compared with that of non-karst areas.

碳酸盐岩的富钙性是导致岩溶动力系统富钙性的根源，在岩溶区大气圈、水圈、生物圈都有明显的表现：大气降水，广西、贵州岩溶地区的雨水中的钙含量达 2.9–6 mg/L，非岩溶地区雨水中钙离子含量常低于 1 mg/L；岩溶水，尤其岩溶地下水中钙离子含量达 50–120 mg/L，而非碳酸盐岩区的外源水中只有几到十几 mg/L；与非岩溶地区的植被相比，生长在岩溶地区的植被具有高的灰分及钙镁含量，而硅铝含量较低。

左上图
Top left

Comparison of major chemical constituents between limestone, dolomite and basalt, granite (with the samples from southwest China).
石灰岩、白云岩与玄武岩、花岗岩中主要化学成分对比（样品来自中国西南）。

左中、左下图
Middle left and bottom left

By comparing sampling 11 types of soils in plant roots that grow in karst and sandstone and shale areas, it was found that the pH value in karst limestone soil is two units higher than that in sandstone and shale areas; while the calcium content in the former soil is 1.5 times that of the latter one.
取 11 种分别生长在岩溶区和砂页岩区的植物根系土壤，检测到岩溶区石灰土的 pH 要高出 2 个单位；前者的钙离子含量是后者的 1.5 倍。

Chemical components of plants living in karst areas and compared with those in non-karst area (percentage in dry plant leaves) 生长在岩溶区的植物化学组成特征（占植物干重的百分比，单位：%）									
Type 类型	Index 指标	Lime 灰分	CaO	MgO	SiO$_2$	Fe$_2$O$_3$	Al$_2$O$_3$	K$_2$O	P$_2$O$_5$
Tropical 热带	Nonggang carbonate rock area 弄岗碳酸盐岩区	7.47	2.88	0.44	0.57	0.03	0.11	0.40	0.28
	Jianfengling granite area 尖峰岭花岗岩区	3.94	0.34	0.22	1.32	#	#	0.64	0.11
Subtropical 亚热带	Maolan carbonate rock area 茂兰碳酸盐岩区	8.71	2.52	0.47	0.74	0.03	0.07	0.89	0.18
	Maolan acidic soil area 茂兰酸性土壤区	5.96	1.49	0.52	1.70	0.53	0.04	0.74	0.53

The table above showing the comparison of the chemical constituents of plant leaves distributed in karst areas and non-karst areas in tropical and subtropical forest areas in China.

上表是分布在中国热带、亚热带森林区岩溶区与非岩溶区植物叶片化学成分的对比。

Vulnerability of Karst Dynamic System
岩溶动力系统的脆弱性

The vulnerability of karst dynamic system is mainly demonstrated from three aspects: deficiency of pedogenic carbonate material leads to soil resource shortage; surface water may easily drop out to underground through sinkholes; vegetation grow under a poor site condition.

岩溶动力系统的脆弱性主要表现在3个方面：碳酸盐岩成土物质先天不足，导致土壤资源短缺；地表水很容易通过落水洞漏失地下；植被的立地条件非常困难。

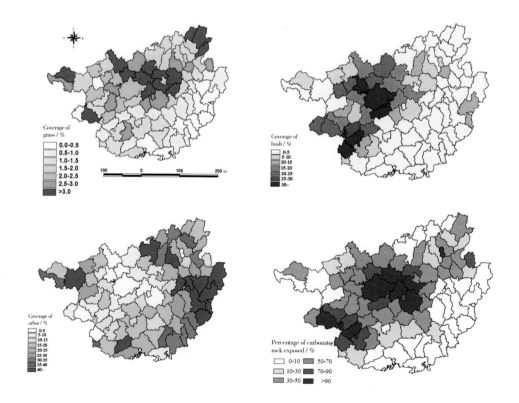

In Guangxi, the karst distribution is negatively correlated with forest cover, while it is positively correlated with the shrub and grass cover.
广西岩溶分布与森林覆盖成负相关，与灌丛、草地覆盖成正相关。

Distributive features of carbonate rock exposure and arbor, bush and grass in each county, GZAR
广西壮族自治区各个县的碳酸盐岩及乔木、灌木和草地的分布特征

上图 / Top
Comparison of soil resources in limestone areas and sandstone and shale areas in Panxian County, Guizhou Province, China.
贵州盘县石灰岩区与砂页岩区土壤资源的对比。

下图 / Bottom
Sinkholes arranged in clusters in Mashan, Guangxi, China.
广西马山成串排列的落水洞。

Characteristics of Karst Dynamic System
岩溶动力系统特点

Glossary of Karstology defines the terms, types and forming mechanism of karst in a standardized way.

《岩溶学词典》对岩溶形态、类型及形成机制进行了规范。

Monograph: Karst Dynamic System in China.

《中国岩溶动力系统》研究专著。

1. The structure and operating patterns of karst dynamic system

(1) The structure of karst dynamic system(KDS)

Carbonate rock provides material basis for KDS, and the classification of carbonate rocks depends on formation environment, mineral composition, and chemical constituents. The karst environment is impacted by limestone formation thickness, occurrence, presence of other rocks and soil cover. A relatively complete karst dynamic system, is usually defined as a karst watershed unit, also involves the impact of lithologies of carbonate rocks on the structure of karst dynamic system.

(2) The operation of karst dynamic system

The operation of Karst dynamic system is subject to multiple factors, such as geological background, weather conditions, hydrological conditions, eco-environment conditions, land use patterns and human activities. Studying of its operation is to design how to reveal its carbon, water, and calcium cycle patterns, as well as the factors and mechanism influencing the karst dynamic system through geochemical method.

2. The type of karst dynamic system and regional distribution pattern

In terms of structure and operating conditions, karst dynamic system is not only restrained by geological conditions, but also is strongly affected by climatic and ecological conditions. Therefore, karst dynamic system has various types and is significant spatial differentiation. As for the cause of formation, some patterns do exist.

3. Basic and applied research on karst dynamic system

(1) Research on karst dynamic system and global climate change

Karst carbon sink effect: While revealing the physical, chemical and biological mechanisms, this study aims to clarify and accurately quantify the karst carbon sink effect in karst process;

Climate change record: True recording of climate change by cave stalagmite in high resolution can help reconstruct the regional climate change pattern and reveal the impact of human activities.

(2) Karst dynamic system and the water cycle and water resources

Hydrological cycle: Karst area has a special surface and underground karst double-layer aquifer. Monitoring and research topics range from rainfall, overland flow, infiltration, hydrology of epikarst zone underground and surface river hydrology, surface evaporation, transpiration and other water transport processes, watershed to be determined, and the water regulation and storage of karst watershed;

Water resource management: exploration, exploitation, utilization and protection of Karst groundwater.

(3) Karst dynamic system and sustainable development

Rocky desertification control and ecological rehabilitation, spatial distributive pattern of soil and water resources in karst dynamic system, and vegetation evolution and migration of nutrients;

Water environment control, occurrence process of droughts and floods in karst areas and their prevention and control; karst groundwater contamination and prevention; karst collapse mechanism and its forecast, prevention and control, etc.;

Karst geomorphical, morphological research and geo-parks in karst area.

1. 岩溶动力系统的结构和运行规律

(1) 岩溶动力系统的结构

碳酸盐岩是岩溶动力系统运行的物质基础，而碳酸盐岩因成因、矿物组成、化学成分而存在差异；不同的岩性在地层中存在的厚度、产状及是否存在其他岩石、土壤覆盖而差异；一个相对完整的岩溶动力系统，通常以岩溶流域为单元，一个流域还涉及碳酸盐岩野外的岩石对岩溶动力系统结构的影响；岩溶形态组合研究方法及岩溶水文流域研究方法是认识岩溶动力系统结构的主要方法。

(2) 岩溶动力系统的运行

岩溶动力系统的运行受到地质背景的影响，但更受气候条件、水文条件、生态环境条件、土地利用方式和人类活动的影响。岩溶动力系统运行的研究就是通过地球化学研究手段，揭示碳、水、钙的运移规律，揭示岩溶动力系统特征塑造的影响因素及机制。

2. 岩溶动力系统的类型和区域分布规律

从岩溶动力系统结构、运行条件看，岩溶动力系统既受地质条件制约，同时又强烈受气候、生态环境的影响，岩溶动力系统既有地带性特征，更有区域性特点，因此，岩溶动力系统具有类型多样、空间分异性显著，从成因上具有一定的规律性可循。

3. 岩溶动力系统的基础研究、应用研究

(1) 岩溶动力系统与全球气候变化研究

岩溶碳汇效应。揭示物理、化学、生物作用机制的同时，澄清其岩溶过程中的碳汇效应及其准确估算；

气候变化记录。洞穴石笋对气候环境变化记录的真实性、高分辨率，有利于对区域气候变化模式的重建；同时有利于揭示人类活动的影响。

(2) 岩溶动力系统与水循环、水资源

水循环过程。岩溶区具有特殊的地表、地下双层岩溶含水介质，从降雨、坡面流、入渗、岩溶表层带水文、地下河水文和地表河流，及地表蒸发、蒸腾等水运移过程的监测研究，流域确定，及流域含水层对水资源的调蓄；

水资源管理。岩溶地下水的探测、开发、利用和保护。

(3) 岩溶动力系统与可持续发展

石漠化综合治理与生态重建，岩溶动力系统中水土资源的空间布局，植物形成演化和营养元素迁移规律；

水环境治理，岩溶区旱涝灾害发生过程与防治；岩溶地下水的污染与防治；岩溶塌陷的机制与预警、预报和防治等；

岩溶地貌、岩溶形态研究与地质公园。

The Research Projects and their Major Progresses
科研项目与主要进展

Carbon cycle: Karst dynamic system and global climate change
碳循环：岩溶动力系统与全球气候变化

UNESCO gives priority to the study on global climate change:
In 2009, UNESCO compiled "The UNESCO climate change initiative" and "The UNESCO strategy for action on climate change". Karst dynamic system is closely associated with global climate change in two aspects: (1) carbonate rocks dissolve and remove atmospheric CO_2 to the hydrosphere, resulting in carbon sink and mitigation of atmospheric greenhouse effect; (2) karst sediments (stalagmites, travertine) extract high-resolution paleoenvironmental change information. Thanks to the support of the Chinese Government, particularly the Ministry of Land and Resources and China Geological Survey, IRCK has carried out research in this area.

联合国教科文组织对全球气候变化的重视：
2009年，联合国教科文组织编著了"The UNESCO climate change initiative"和"The UNESCO strategy for action on climate change"，岩溶动力系统从两个方面与全球气候变化发生密切的联系：碳酸盐岩溶解消耗将大气CO_2转移到水圈中，产生碳汇，缓解大气温室效应；岩溶沉积物（石笋、钙华）提取高分辨率古环境变化信息。在中国政府，尤其国土资源部、中国地质调查局的支持下，国际岩溶研究中心开展了相关的研究。

Some projects related to global climate change:
(1) China Geological Survey Project "Karst dynamic system and carbon cycle" (1212010911062), 2009-2010, Yuan Daoxian, Cao Jianhua;
(2) China Geological Survey Project "Karst carbon sink process and its effect in China" (1212011087120), 2010.1-2012.12, Cao Jianhua;
(3) China Geological Survey Project "Karst carbon sink dynamic assessment in China" (1212011087121), 2010.1-2012.12, He Shiyi;
(4) China Geological Survey Project "Comprehensive study on geological carbon sink potential" (1212011087119), 2010.1-2012.12, Yuan Daoxian, Zhang Cheng;
(5) China Geological Survey Project "Investigation on the background and condition for karst carbon sink in China", 2010.1-2012.12, Qin Xiaoqun;
(6) Special Fund of the Ministry of Land and Resources for Scientific Research in Public Interest "Dynamic monitoring and evaluation for carbon sink in typical karst dynamic systems in China" (201111022), 2011.1-2013.12, Zhang Cheng;

(7) Special Fund of the Ministry of Land and Resources for Scientific Research in Public Interest "Survey and dynamic monitoring methods and training for karst carbon sink worldwide" (201211092), 2012.1–2014.12, Cao Jianhua;

(8) Outstanding Youth Project of the Ministry of Land and Resources for Scientific Research in Public Interest "Carbon cycle process and effect in typical reservoirs in karst areas", 2013.1.1–2015.12.31, Pu Junbing;

(9) China Geological Survey Project "Research on abrupt climate change by high resolution records in stalagmite" (1212011087115), 2010.1–2012.12, Zhang Meiliang;

(10) Chinese National Natural Scientific Foundation Project "Climate changes recorded by stalagmite over the last 500,000–150,000 years in Southwest China and their global significance" (40772216), 2008.1–2010.12, Zhang Meiliang;

(11) Youth Science Fund Chinese National Natural Scientific Foundation Project "Internal structure characteristics of MIS5a/4 abrupt climate change recorded by cave stalagmite in high resolution in Southwest China" (40802042), 2009.1–2011.12, Zhu Xiaoyan.

近年来与碳循环相关的科研项目：

[1] 中国地质调查项目"岩溶动力系统与碳循环"（1212010911062），2009–2010，袁道先、曹建华；

[2] 中国地质调查项目"中国岩溶碳汇过程与效应研究"（1212011087120），2010.1–2012.12，曹建华；

[3] 中国地质调查项目"中国岩溶碳汇动态评价"（1212011087121），2010.1–2012.12，何师意；

[4] 中国地质调查项目"地质碳汇潜力综合研究"（1212011087119），2010.1–2012.12，袁道先、章程；

[5] 中国地质调查项目"中国岩溶碳汇典型流域调查"，2010.1–2012.12，覃小群；

[6] 国土资源部公益性行业专项"我国典型岩溶动力系统碳汇动态监测与评价"（201111022），2011.1–2013.12，章程；

[7] 国土资源部公益性行业专项"全球岩溶碳汇调研及动态监测方法与培训"（201211092），2012.1–2014.12，曹建华；

[8] 国土资源部公益性行业优秀青年项目"岩溶区典型水库岩溶作用碳循环过程及效应研究"，2013.1.1–2015.12.31，蒲俊兵；

[9] 中国地质调查项目"高分辨率洞穴石笋记录的气候突变过程及其规律性研究"（水[2011]01-14-03，1212011087115），2010.1–2012.12，张美良；

[10] 国家自然科学基金项目"中国西南50万–15万年来石笋记录的气候事件及全球意义"（40772216），2008.1–2010.12，张美良；

[11] 国家自然科学青年基金"中国西南高分辨率洞穴石笋记录的MIS5a/4气候突变的内部结构特征"（40802042），2009.1–2011.12，朱晓燕。

The study of IGCP 379 "Karst Processes and the Carbon Cycle" (1995-1999) found that, by estimating the global carbonate dissolution, the annual consumption of atmospheric CO_2 is 0.6 PgC, which has been validated. However, there is still controversy over this finding in the academic community. The traditional view held that CO_2 is dissolved and consumed by carbonate rocks on continent, in the form of inorganic carbon in the water body (HCO_3^-), migrateing to the ocean along with the river, generateing carbonate deposits in the ocean; CO_2 returns to the atmosphere, and therefore, carbonate dissolution process does not produce a net carbon sink. However, the recent studies showed that the dissolution of carbonate leads to the concentration of inorganic carbon in karst water with an order of magnitude higher than that in granite and sandstone and shale areas, high concentrations of HCO_3^- stimulate photosynthesis of aquatic plants (photosynthesis of aquatic plants is constrained by the amount of carbon source), convert inorganic carbon into organic carbon, and enter into the aquatic biosphere. Some organic carbon is deposited in the terrestrial water bodies or wetlands. From small basin scale to global scale, carbon flux dissolved by carbonate has the same order of magnitude as the carbon sink flux produced by terrestrial vegetation and soil.

The latest results indicate that:

(1) karst carbon sink process is closely related to bio-organic processes in the earth surface system. It not only involves the traditional long-period geological processes, but also process at short time scale. Carbon sink flux in karst can be comparable to that in terrestrial ecosystems;

(2) the process of karst carbon sink is subject to human activities, such as land use pattern and land cover change as well as control of rocky desertification and ecological restoration;

(3) aquatic plants play an important role in maintaining the stability of karst carbon sinks, and substantive progress has been achieved.

IGCP 379 "岩溶作用与碳循环"（1995-1999）的研究结果显示，通过估算全球因碳酸盐岩溶解每年可消耗大气 CO_2 0.6 PgC，这一结果后来得到验证。但在学术界还是存在争议，争议的焦点是，传统的观点认为陆地上碳酸盐岩溶解消耗的 CO_2，在水体中以无机碳（HCO_3^-）的形式，并随着河流迁移到海洋中，在海洋中发生碳酸盐沉积，CO_2 重新回到大气，碳酸盐岩溶解过程不产生净碳汇。最近的研究结果显示，碳酸盐岩的溶解导致岩溶水体中无机碳浓度比花岗岩、砂页岩地区水体中的高一个数量级，主要以 HCO_3^- 形式存在，且高浓度 HCO_3^- 的岩溶水，刺激水生植物的光合作用（水生植物的光合作用受到碳源多少的制约），将无机碳转化为有机碳，而进入水生生物圈，部分有机碳则在陆地水域或湿地中沉积；从小流域到全球尺度，碳酸盐岩溶解产生的碳汇通量与陆地地表植被和土壤产生的碳汇通量处于同一个数量级。

通过最新的研究，主要取得以下主要进展：

1. 岩溶碳汇过程与地球表层系统中生物有机过程密切相关，不仅有传统长周期的地质过程，更有短时间尺度的碳汇过程；产生的碳汇通量可与陆地生态过程碳汇通量相比；

2. 岩溶碳汇过程受到人为活动的影响，如土地利用方式的改变、石漠化治理产生的碳汇效应；

3. 水生植物对岩溶碳汇的稳定性具有重要作用，并取得实质性进展。

Karst Dynamic System and Carbon Sink Process and Effect
岩溶动力系统与碳汇过程与效应

The annual carbon flux emitted by limestone soil respiration in Guilin Maocun karst underground water catchment is found 25% less than that emitted by red soil in sandstone and shale area
桂林毛村地下河流域中岩溶区石灰土土壤呼吸排放 CO_2 的年通量比砂页岩区红壤的少 25%

Elucidating soil carbon sinks in karst areas have been a major research focus, while the relationship between karst soil carbon transfer and karst soil carbon sinks is unclear. A comparative study of soil respiration rates, $\delta^{13}C$ values of CO_2 from soil respiration, and CO_2 concentrations in forest soils was conducted from September 2006 to August 2008 between karst and clasolite areas of the Maocun karst underground river catchment. The main results are as follows: ① the soil respiration rate in limestone soil from the karst area is apparently lower than that in red soil from the clasolite area. The soil respiration rate in the karst areas varied from 23.12~271.26 $mgC/(m^2 \cdot h)$, with an average of 111.57 $mgC/(m^2 \cdot h)$, while the soil respiration rate in the clasolite areas varied from 51.60~326.28 $mgC/(m^2 \cdot h)$, with an average of 148.99 $mgC/(m^2 \cdot h)$. Taking the averages into account, the soil respiration rate in limestone soil was 25.12% less than that in red soil; ② the $\delta^{13}C$ values of soil respiration in the karst and clasolite areas were −29.35‰~−18.26‰ and −29.21‰~−22.60‰, respectively, with respective mean values of −22.68‰ and −26.21‰. The $\delta^{13}C$ value of soil respiration in the karst areas was greater than that in the clasolite areas; ③ the CO_2 concentration of limestone soil profiles had a bi-directional gradient, which was more obvious in seasons with good hydro-thermal conditions. In contrast, the CO_2 concentration of red soil profile had a uni-directional gradient, that is to say, the deeper the red soil layer, the higher the CO_2 concentration. Taking into account the mean values of CO_2 concentration, the CO_2 concentration of limestone soil ranged between 0.05%~0.60%, with an annual average of 0.25%. The CO_2 concentration of red soil ranged between 0.05%~1.09%, with an annual average of 0.57%, which indicated that the lower soil CO_2 could be consumed and absorbed by carbonate rock dissolution at the soil/rock interface in the karst areas. In other words, karst processes in soil represented one of the carbon sinks.

桂林毛村岩溶地下河流域位于桂林市东南 30 km 的潮田毛村，流域面积约 10 km²，为开展岩溶区、碎屑岩区土壤碳迁移对比研究，选择具代表性的林下棕色石灰土、红壤剖面各一个开展以月为观测周期的土壤呼吸排放 CO_2 速率、同位素动态变化及其土壤剖面不同层位 CO_2 浓度时空分布的动态观测，进而分析岩溶动力系统中土壤碳迁移特征及岩溶碳汇作用机制。结果显示：① 岩溶区石灰土的土壤呼吸排放 CO_2 速率明显低于碎屑岩区红壤的，岩溶区土壤呼吸速率的变化幅度为 23.12 ~ 271.26 mgC/(m²·h)；碎屑岩土壤呼吸速率的变化幅度为 51.60 ~ 326.28 mgC/(m²·h)，如以年平均值计算，则岩溶区土壤呼吸排放 CO_2 的量要比碎屑岩区红壤少 25.12%；② 岩溶区石灰土土壤呼吸排放 CO_2 的 $\delta^{13}C$ 值比碎屑岩区红壤的偏重，岩溶区土壤呼吸排放 CO_2 的 $\delta^{13}C$ 值为 −29.35‰ ~ −18.26‰，平均为 −22.68‰，碎屑岩区为 −29.21‰ ~ −22.60‰，平均为 −26.21‰；③ 岩溶区石灰土剖面中 CO_2 浓度出现双向梯度，且水热条件良好的季节双向梯度表现越明显，而碎屑岩区红壤剖面中则出现随土壤层深度的增加，土壤 CO_2 浓度增加的一向梯度；这意味着岩溶区土—岩界面石灰岩的溶解消耗吸收土壤层 CO_2，即土壤中岩溶作用产生碳汇的过程。

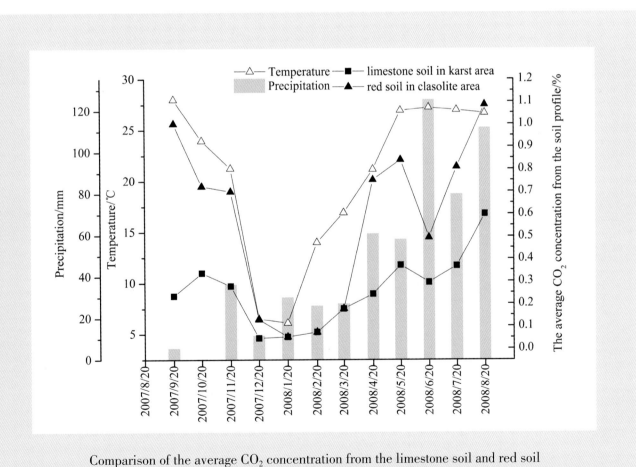

Comparison of the average CO_2 concentration from the limestone soil and red soil
岩溶区石灰土与碎屑岩区红壤土壤呼吸排碳动态对比

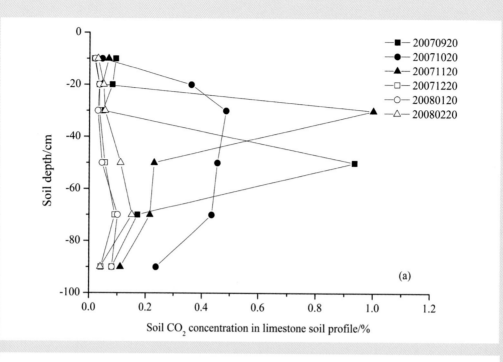

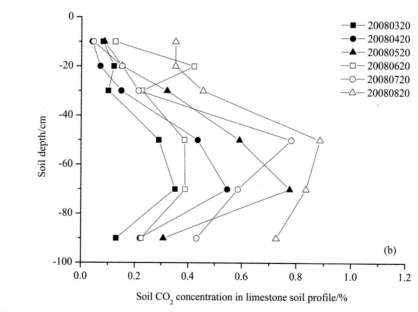

Change of soil CO$_2$ concentration in limestone soil profile
岩溶区石灰土土壤剖面 CO$_2$ 浓度动态变化

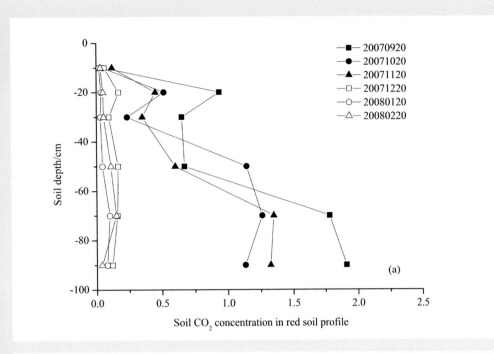
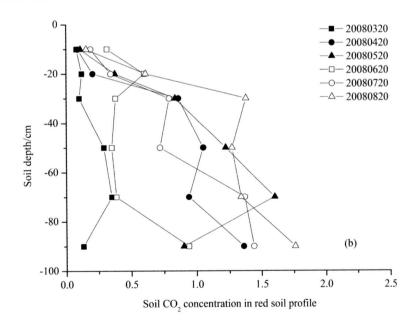

Change of soil CO_2 concentration in red soil profile
碎屑岩区红壤土壤剖面 CO_2 浓度动态变化

The Soil Carbonic Anhydrase (CA) Activity Affected by the Development of Plant Roots (Soil Biogenic Activity)
土壤碳酸酐酶的活性受到植物根系发育程度的影响

CA as the enzyme can efficiently catalyze the reaction: $CO_2+H_2O \leftrightarrow HCO_3^- + H^+$ and the CA can be widely found in plants, animals, micro-organisms and debris. The sampling test has been done in different land-use patterns and soil depths in karst area to study the CA activity.

The results (shown in the below four figures) indicate that the CA activity has no obvious variation regularity in different soil depths for the strong taproots in forest land. However, in the shrub land, the CA activity increased with the depth of soil due to widely developed roots. On the contrary, in the abandoned farm land, the CA is more active at the superficial soil layer.

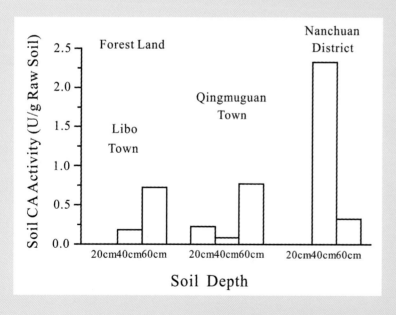
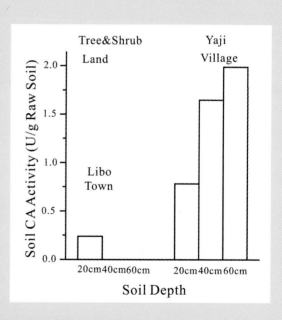

Test results and comparison of the CA in different land-use patterns and soil depths in karst area
不同土地利用方式、不同土壤深度碳酸酐酶的检测结果及对比

碳酸酐酶具有高效快速催化 CO_2 和 HCO_3^- 之间的相互转化的功能，广泛来源于植物、动物、微生物体及残体。本项目通过不同地区、不同土地利用方式和土壤深度进行土壤碳酸酐酶 (CA) 的检测。

检测的结果（如下图）显示，林地由于主根系发达，不同深度土壤碳酸酐酶的活性变化规律不明显；而乔灌林、灌木林根系繁盛，随深度有明显变化，而碳酸酐酶活性随深度增加而增加；对于弃耕地而言，碳酸酐酶的活性则主要集中在浅表层。这意味着土壤碳酸酐酶的活性受到植被根系的影响，进而影响土壤环境中的碳迁移和转化。

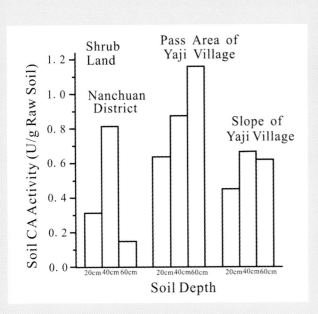
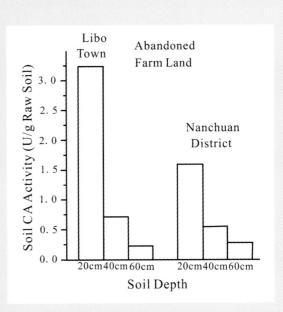

Test results and comparison of the CA in different land-use patterns and soil depths in karst area
不同土地利用方式、不同土壤深度碳酸酐酶的检测结果及对比

Carbonate rock dissolution rates in different land cover patterns and their carbon sink effect
不同土地利用下的碳酸盐岩溶蚀速率及它们的碳汇效应

Researching on karst processes with various land use and land cover patterns can contribute to the estimation of karst carbon sink, and the accurate evaluation of carbon source/sink in karst area as well. Using limestone standard tablets, the carbonate dissolution rates in the soil under different land use patterns in two typical karst catchments (i.e., Nongla peak-cluster depression in Guangxi and Jinfo Mountain karst valley in Chongqing) were studied. The results showed that there is a sharp distinction in dissolution rates in the soil under different land cover patterns, with generally higher rate recorded in the forest. The mean values of tablets weight loss in cultivated land, shrub, secondary forest, grassland and primary forest are 4.02 t/(km$^2 \cdot$ a), 7.0 t/(km$^2 \cdot$ a), 40.0 t/(km$^2 \cdot$ a), 20.0 t/(km$^2 \cdot$ a), 63.5 t/(km$^2 \cdot$ a) respectively. Accordingly, in the proceeding of karst carbon sink estimation in regional scale, land cover difference should be taken into account, as other impacting factors than climate, hydraulic and geological conditions. Vegetation evolution could improve karst carbon sink remarkably, carbon sink resulted from karst processes under the soil is three times higher in primary forest than that in secondary forest, and it is nine times higher than that in shrub. That is to say, carbon sink resulting from karst processes will respectively increase 5.71–7.02 t/(km$^2 \cdot$ a) and 24.86–26.17 t/(km$^2 \cdot$ a) from cultivated land or shrub to secondary forest and to primary forest. Carbon sink of terrestrial ecosystem increases with vegetation positive evolution or reforestation, and the similar process caused by karst dissolutional denudation will simultaneously occur in the soil.

不同土地利用下的岩溶作用研究不仅关系到区域岩溶碳汇估算，也关系到岩溶区陆地碳源/汇的准确评估。利用标准溶蚀试片法研究了广西弄拉峰丛洼地和重庆金佛山岩溶谷地2个典型岩溶动力系统内不同土地利用下的土下溶蚀速率。结果显示，不同土地利用下的土下溶蚀速率差异较明显，耕地、灌丛、次生林、草地、原始林平均值分别为4.02 t/（km$^2 \cdot$ a）、7.0 t/（km$^2 \cdot$ a）、40.0 t/（km$^2 \cdot$ a）、20.0 t/（km$^2 \cdot$ a）和63.5 t/（km$^2 \cdot$ a）。因此，在进行区域尺度岩溶作用碳汇估算时，除了考虑气候、水文、地质等条件外，还必须考虑土地利用类型的差异。植被的正向演替对岩溶碳汇有显著的促进作用，原始林地土下岩溶作用碳汇量是次生林地的3倍，灌丛的9倍，也就是说，从耕地或灌丛演化到次生林地，由岩溶作用产生的碳汇可提高5.71-7.02 t/km$^2 \cdot$ a，若演化到原始林地则达24.86-26.17 t/km$^2 \cdot$ a。岩溶区地表森林系统的增汇过程发生的同时，地下也同步发生着类似的增汇过程。

The carbonate dissolution rates in the soil under different land use patterns in two typical karst catchments
两个典型岩溶动力系统内不同土地利用下的土下溶蚀速率

Location 地点	Land and land cover 土地利用	Dissolution rate/(t/(km^2·a)) 单位面积溶蚀量	Unit area CO$_2$ consumption /(t/(km^2·a)) 单位面积 CO$_2$ 消耗量	Carbon sink /(t/(km^2·a)) 碳汇
Nongla, Mashan County, Guangxi 广西马山弄拉	Farmland in Nongtuan 弄团耕地	4.02	1.769	0.482
	Shrub in Dongwang 东旺灌丛	0.51	0.224	0.061
	Lan Dian Tang Forest 蓝电堂林地	19.97	8.787	2.396
	Lan Dian Tang Garden 蓝电堂园地	32.97	14.507	3.956
Jinfo Mountain, Chongqing 重庆南川金佛山	Rocky desertification area 石漠地	10.38	4.567	1.246
	Bi Tan Quan Forest 碧潭泉林地	20.0	8.80	2.40
	Bi Tan Quan Shrub 碧潭泉灌丛	7.0	3.08	0.84
	Shui Fang Quan Forest 水房泉林地	63.5	27.94	7.62
	Shui Fang Quan Grass 水房泉草地	40.0	17.60	4.80

Comparison of soil organic carbon stability in karst and non-karst area
岩溶地区及非岩溶地区土壤有机碳稳定性对比

Using the soil samples of 0–20 cm and 20–50 cm in depth from dry land, paddy field, shrub land and woodland in the karst and non-karst areas of Maocun Village in Guilin, the method of potassium dichromate-sulphuric acid oxidation was taken to test the soil organic content (SOC), while 333.33 mmol/L potassium permanganate oxidation method was employed to test the content of the liable organic carbon (LOC).

Soil carbon pool's liability (L) is the liable organic carbon (LOC) in the soil divided by non-liable organic carbon (NLOC) in the soil, which can be expressed as:

$$\text{Soil carbon pool's liability (Lability, L)} = \frac{\text{Liable organic carbon (LOC)}}{\text{Non-liable organic carbon (NLOC)}}$$

Where NLOC=SOC − LOC.

The test results showed that apart from liability of the 0–20cm depth of limestone soil which is higher than that of red soil, the liability of other land use patterns in red soil is generally higher than that in limestone soil. This, to some extent, indicated that the LOC in red soil has a larger proportion of the total organic carbon, and its carbon pool is unstable, while the organic carbon pool in limestone soil is relatively stable.

以桂林毛村岩溶区和非岩溶区的旱地、水田、灌丛、林地 0–20 cm、20–50 cm 的土壤样为研究对象，用重铬酸钾–硫酸氧化法测定其土壤有机碳 (SOC) 含量，用 333.33 mmol/L 的高锰酸钾氧化法测定土壤易氧化有机碳 (LOC) 组分含量。

同时用土壤碳库活度（Liability, L）来表示其不稳定性，它等于土壤中的易氧化有机碳与非易氧化有机碳（Non-liable organic carbon，NLOC）之比：

$$\text{碳库活度（Liability, L）} = \frac{\text{样品中活性有机碳（LOC）}}{\text{样品中非活性有机碳（NLOC）}}$$

土壤样品中非活性有机碳含量 (NLOC)= 土壤有机碳含量 (SOC)− 土壤易氧化有机碳 (LOC) 组分含量。

其检测结果显示，除旱地 0–20cm 土层石灰土的碳库活度大于酸性土外，其余各土地利用方式的碳库活度均为酸性土＞石灰土。这在一定程度上说明酸性土的活性有机碳占总有机碳的比例较高，碳库不稳定；而石灰土中的有机碳库具有相对稳定性。

Content distributions of SOC under different land use patterns in karst area and clastic area
岩溶区与碎屑岩区不同土地利用方式土壤总有机碳含量分布

Land use 土地利用类型	Soil thickness 土层深度	Limestone soil /(g/kg) 岩溶区土壤	Average/(g/kg) 平均含量	Red soil /(g/kg) 碎屑岩区土壤	Average/(g/kg) 平均含量
Dry land 旱地	0–20 cm 20–50 cm	17.75 14.92	16.34	12.55 11.97	12.26
Paddy field 水田	0–20 cm 20–50 cm	25.92 8.80	17.36	19.85 3.70	11.78
Shrub 灌丛	0–20 cm	39.84	39.84	18.58	18.58
Forest 林地	0–20 cm 20–50 cm	31.98 13.91	22.95	27.57 15.75	21.66
Total organic carbon 总有机碳含量			24.12		16.07

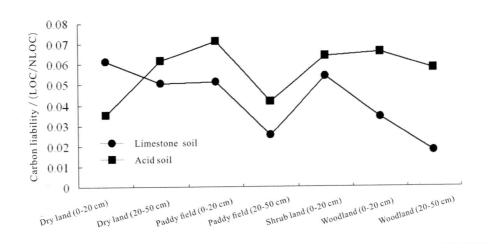

Soil carbon pools liability under different land use patterns
不同土地利用方式下的土壤碳库活度

Allogenic water can increase karst carbon sink flux
外源水可增加岩溶碳汇量

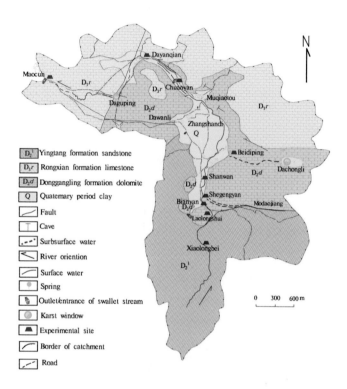

Maocun subterranean stream
and monitoring sites distribution, Guilin
桂林毛村流域及监测点分布图

The impact of allogenic water on karst carbon sink flux in the Maocun subterranean stream in Guilin was studied. Eight sites were selected for comparative study, namely, Xiaolongbei (XLB), Bianyan (BY), Shegengyan (SGY), Shanwan (SW), Chuanyan (CY), Dayanqian (DYQ), Maocun (MC) and Beidiping (BDP). Among these sites, XLB is a surface water site in the sandstone and shale area, and others are karst springs. The annual mean HCO_3^- concentrations are BDP(4.65 mmol/L) > MC(3.67 mmol/L) > DYQ (3.61 mmol/L) > CY(3.58 mmol/L) > SW(2.42 mmol/L) > SGY (2.1 mmol/L) > BY (1.59 mmol/L) > XLB(0.2 mmol/L). The highest concentration of HCO_3^- is in BDP ranged from 3.5 to 5 mmol/L pare limestone in recharge area. Comparatively, the lowest one is in XLB ranged from 0.1 to 0.4 mmol/L due to rainfall input directly and location in sandstone area. The former one is roughly 10 times higher than the latter one.

The above figures show that after allogenic water is mixed with karst water, the SIC and SID indexes will drop, which can help stabilize the HCO_3^- in water, increase erosion of the water, increase the dissolution of carbonate, consume more CO_2 from atmosphere and soil, and enhance potential of carbon sink in karst area.

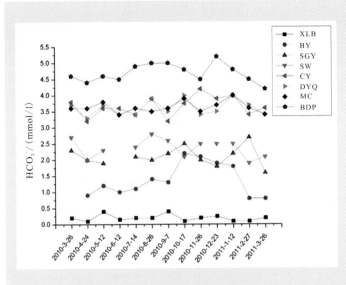 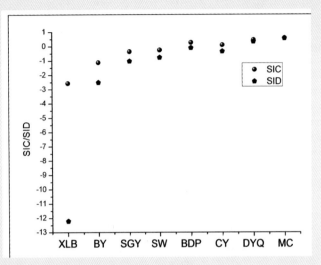

Dynamic characteristics of HCO$_3^-$ in study area
研究区 HCO$_3^-$ 的动态变化特征

Saturation index of calcite in study area
研究区方解石饱和指数

以桂林毛村岩溶地下河有砂页岩地区外源水的输入，小龙背为代表性的砂页岩区小溪流，背地坪为代表性的岩溶泉，因此，从 HCO$_3^-$ 浓度年平均值变化看：背地坪（4.65 mmol/L）＞毛村（3.67 mmol/L）＞大岩前（3.61 mmol/L）＞穿岩（3.58 mmol/L）＞山湾（2.42 mmol/L）＞社更岩（2.1 mmol/L）＞扁岩（1.59 mmol/L）＞小龙背（0.2 mmol/L），穿岩、大岩前与毛村地下河出口的 HCO$_3^-$ 浓度相差不大（左图）。背地坪因没有外源水补给的影响，其 HCO$_3^-$ 浓度最大，变化范围为 3.5–5 mmol/L，而小龙背则全由大气降水直接补给且位于碎屑岩区，所以其 HCO$_3^-$ 浓度最小，变化范围为 0.1–0.4 mmol/L，两地 HCO$_3^-$ 浓度比较，前者为后者的 10 倍。

这意味着外源水与岩溶水混合后，能降低其 SIc（方解石饱和指数）与 SId（白云石饱和指数），不仅有利于水体中碳的稳定，同时提高水体的侵蚀性，增加碳酸盐岩的溶解，消耗更多大气/土壤中的 CO_2，提高了岩溶区的碳汇潜力。

9% of HCO$_3^-$ in karst water is converted into organic carbon when the ground water coming out and flowing 1350 m distance, a subterranean stream in Guancun, Guangxi
广西官村岩溶地下河出流 1350 m 后，岩溶水体中 HCO$_3^-$ 有 9% 转化为有机碳

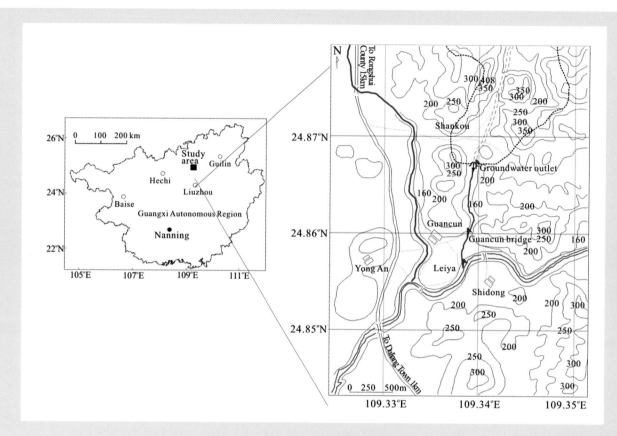

Three monitoring sites in Guancun
官村的 3 个观测点的分布图

Guancun underground stream river, in Rongshui County of Guangxi, China, three monitoring sites were selected when the groundwater coming out: (1) at the outlet; (2) Guancun Bridge; (3) Leiya Village, about 1350 m far away from outlet. And one week of monitoring (one group of data every five minutes) and sampling (per hour) were conducted to examine the dynamic variation of indicators in the water (pH, electrical conductivity, soluble oxygen, HCO$_3^-$ and $\delta^{13}C_{DIC}$) between day and night, and the loss of DIC at the monitored river section was also estimated. The results indicated that the chemical, isotope and other indicators at the subterranean stream outlet in Guancun kept relatively stable without any change between day and night; while the indicators at the Guancun Bridge and Leiya Village

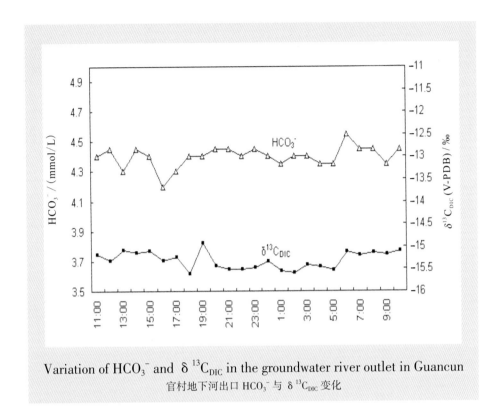

Variation of HCO_3^- and $\delta^{13}C_{DIC}$ in the groundwater river outlet in Guancun

官村地下河出口 HCO_3^- 与 $\delta^{13}C_{DIC}$ 变化

demonstrated significant change between day and night. The calculation of less of HCO_3^-, depending on the mass balance, showed that the output value of inorganic carbon at the monitored river section is lower than that of the subterranean stream outlet. Bicarbonate gradually declines along the way. After the flow path of 1350 m, the HCO_3^- in karst water is lost by 94.9 kg each day, accounting for 9% of the total inorganic carbon.

以广西融水县官村地下河出口地表河段为例，通过一周的连续加密监测(5分钟一组数据)和取样(每小时)，研究了其水体中各监测指标(pH、电导率、溶解氧、HCO_3^-及$\delta^{13}C_{DIC}$)昼夜动态变化，并估算了监测河段 DIC 的丢失量。研究结果表明，官村地下河出口水化学、同位素等指标相对稳定，无昼夜变化；下游官村桥（800 m 处）和雷崖村汇合处（1350 m 处），则各项指标变幅增大，出现显著的昼夜变化。质量平衡计算结果表明：监测段河流无机碳输出值低于地下河出口输出值，重碳酸盐沿途逐渐减少，1350 m 流程后，岩溶水体中每天 HCO_3^- 流失量估算值为 94.9 kg，占无机碳总量的 9%。

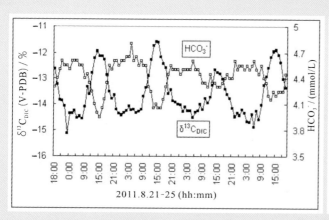
Guancun Bridge HCO_3^- and $\delta^{13}C_{DIC}$ variation

官村桥 HCO_3^- 与 $\delta^{13}C_{DIC}$ 变化

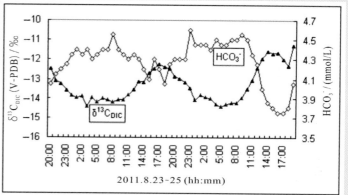
Leiya Village confluence HCO_3^- and $\delta^{13}C_{DIC}$ variation

雷崖村汇合处 HCO_3^- 与 $\delta^{13}C_{DIC}$ 变化

Native chlorella vulgaris enhancing the conversion rate of HCO_3^- in karst water
土著小球藻可提高岩溶水体中 HCO_3^- 转化效率

Exogenous chlorella vulgaris and native chlorella vulgaris which from karst areas were selected. The comparison study on the utilization of Ca^{2+} and HCO_3^- in karst water by chlorella vulgaris of two different origins in a closed system, and chlorella vulgaris cell numbers and the utilization rate of Ca^{2+} and HCO_3^- and the pH value change were tested. The results showed that the native chlorella vulgaris utilized more Ca^{2+} and HCO_3^- than exogenous chlorella vulgaris, while exogenous chlorella vulgaris utilized more Ca^{2+} than native chlorella vulgaris, but they utilized the same amount of HCO_3^-. In addition, exogenous chlorella vulgaris can form $CaCO_3^-$ rich sediment in the form of extracellular crystal, but that is not for native chlorella vulgaris. Furthermore, the pH value changes in the closed system revealed that both algae utilize the dissolved carbon dioxide as photosynthetic carbon source and make use of HCO_3^-. Exogenous chlorella vulgaris can absorb 26.3% Ca^{2+} and 29.6% HCO_3^- from the karst water, and native chlorella vulgaris makes use of 42.1% Ca^{2+} and 40.6% HCO_3^-. As a primary producer in the food chain, the two kinds of aquatic algae transform HCO_3^- and take it into the ecological system, serving as a net carbon sink.

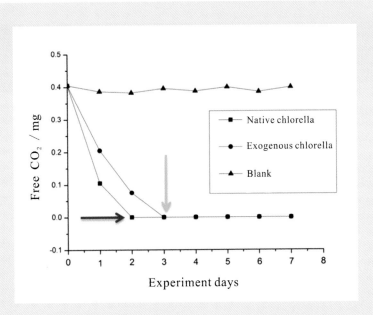

The change curve of free CO_2 in the closed system

封闭体系中游离 CO_2 变化曲线

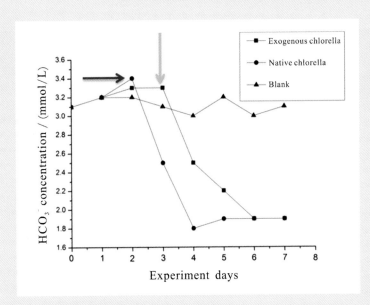

The alteration curve of HCO_3^- concentration in the closed system

相同培养条件下岩溶水中 HCO_3^- 变化曲线

以外源小球藻和岩溶区筛选出的土著小球藻为研究对象,在封闭体系中比较研究了两种不同来源小球藻对典型岩溶水中 Ca^{2+}、HCO_3^- 的利用、藻细胞数量与其对 Ca^{2+}、HCO_3^- 的利用率的关系和体系 pH 的变化。结果表明,土著小球藻利用 Ca^{2+}、HCO_3^- 的能力强于外源小球藻,但外源小球藻对 Ca^{2+} 的利用量高于土著小球藻,而两者对 HCO_3^- 的利用量相同,并且外源小球藻能够以胞外 $CaCO_3$ 形式产生沉淀,而土著藻则不能形成沉淀。其次两体系中 pH 的变化显示,两种小球藻光合作用都是先以水体中 CO_2 为光合作用碳源,然后利用 HCO_3^-。外源小球藻能将岩溶水中 29.648% 的 HCO_3^- 吸收,而土著藻能将 40.625% 的 HCO_3^- 通过其在食物链中的初级生产地位将岩溶碳汇转化进入到生态系统,表现为净碳汇效应。

Carbon sink intensity from karst process in Pearl River Basin estimation to be 40 t $CO_2/km^2 \cdot a$
珠江流域岩溶碳汇强度估算达到 40 t $CO_2/km^2 \cdot a$

The Pearl River Basin is an important subtropical region in China characterized by karst. It is one of the three largest river basins in China, with a river basin area of 452,600 km². It has a karst area of 181,800 km², representing 40.1% of the river basin recharge area.

We analyzed comprehensively the factors impacting karstification and the carbon sink, collected existing monitoring data, and established a regression equation incorporating corrosion rate, annual precipitation, soil respiration rate and net primary productivity from typical observation sites. Arcview 3.3 software was also used to estimate spatially the atmospheric CO_2 sink flux in the Basin's karst region by combining the distribution of carbonate rock categories. The annual CO_2 consumption due to carbonate rock corrosion was estimated to be 1.54×10^7 t $CaCO_3$/ a, equal to 1.85×10^6 t C/ a or 42.70 t $CO_2/km^2 \cdot a$.

珠江流域是我国南方亚热带湿润地区重要的、以岩溶发布为主要特征的、我国三大流域之一，流域面积 45.26 万 km²，岩溶地区分布的面积达 18.18 万 km²，岩溶面积占流域面积的 40.1%。

Estimation of carbon sink flux by karstification in the Pearl River Basin 珠江盆地岩溶碳汇通量的估算				
Lithology 岩性	Area/km² 面积/km²	Sepecific corrodililily 溶蚀度	Annual corrosion /(t $CaCO_3$ a^{-1}) 年溶蚀速率/(t $CaCO_3$ a^{-1})	Carbon equivalent /(t Ca^{-1}) 碳平衡/(t Ca^{-1})
Dolomite 白云岩	4 238.47	0.505	246 676.23	29 601.15
Dolomitic limestone 白云质灰岩	6 604.49	0.833	633 997.49	76 079.70
Impure dolomite 含杂质白云岩	2 509.31	0.767	221 795.60	26 615.47
Impure limestone 含杂质石灰岩	51 346.28	0.767	4 536 675.31	544 401.04
Limestone 石灰岩	63 198.02	0.965	7 028 036.93	843 364.43
Cacareous dolomite 石灰质白云岩	30 543.74	0.770	2 710 292.66	325 235.12
Total amount 总量	158 440.31		15 377 474.22	1 845 296.91

在综合调查珠江地质、水文、植被覆盖等背景条件下，考虑影响岩溶作用及碳汇因子的基础上，收集已有的数据，以典型点的溶蚀速率、年降雨量、土壤呼吸速率和 NPP 建立回归方程，并以 GIS 为研究平台，结合区内碳酸盐岩类型的分布，估算研究区因岩溶作用对大气 CO_2 汇的通量，试图从原位典型点的监测数据，探索区域尺度岩溶作用碳汇估算方法、由点到面的技术途径。估算结果显示，珠江流域年溶蚀量为 1.54×10^7 t$CaCO_3$/a，折合碳为 1.85×10^6 tC/a，相当于 42.70 t $CO_2/km^2 \cdot a$。

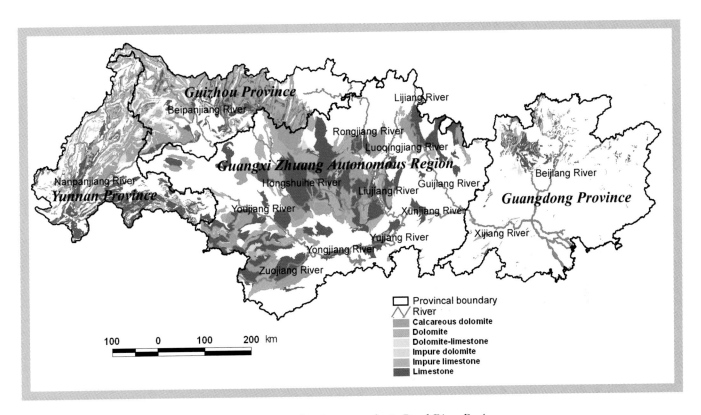

Different types of carbonate rocks in Pearl River Basin
珠江流域不同碳酸盐岩类型分布图

Karst carbon sink flux in the same order magnitude as that of terrestrial ecosystem
岩溶碳汇通量与陆地生态碳汇通量处于同一数量级

Carbonate rocks are material basis for karst dynamic system. Taking up 9.3%–15.9% of the continent area, the carbonate rocks dissolution is the principal source and transfer most of the DIC from terrestrial water to ocean. The carbon sink flux generated by carbonate dissolution contributes 37%–51.4% of continent weathering carbon sink. The data on global atmospheric/soil CO_2 fluxes dissolved and consumed by carbonates vary with different researchers and methods, ranging from 0.1 to 0.6 PgC/a. Taking the intermediate value of 0.3 PgC/a, global terrestrial ecosystem photosynthesis will produce an annual net carbon flux of 1.7±0.5 PgC/a, while the global soil organic carbon pool under rational management can produce a net carbon flux of 0.4–1.2 PgC/a. Therefore, carbon sink flux generated by karst process accounts for about 17.65% of the total carbon flux in terrestrial forest ecosystems and 37.5% of total soil carbon flux respectively.

If the atmospheric / soil CO_2 consumed by carbonate rocks in China is estimated to be 0.016 PgC/a, it is found that in 1981 to 2000, net annual carbon sink volume of forests ecosystem at national scale reached 0.075 PgC/a, and the amount of soil carbon sink 0.04–0.07 PgC/a. If so, karst carbon sink flux at national scale accounted for 21.3% in forest ecosystem net carbon sink, and 22.9%–40% in soil carbon sink.

岩溶动力系统以碳酸盐岩为物质基础，占陆地面积9.3%–15.9%的碳酸盐岩溶解构成陆地水域向海洋输送DIC的主体，碳酸盐岩溶解产生的碳汇通量对陆地风化碳汇的贡献达到37%–51.4%；全球碳酸盐岩溶解消耗大气/土壤CO_2的通量因不同研究者和不同方法有所差异，为0.1–0.6 PgC/a，如取中间值0.3 PgC/a；全球陆地生态系统植物的光合作用将消耗大气CO_2，陆地森林生态系统将产生1.7±0.5 PgC/a的净碳通量，而全球土壤有机碳库在合理的管理下可产生净碳通量0.4–1.2 PgC/a；为此，岩溶作用产生的碳汇通量约占陆地森林碳汇通量的17.65%、土壤碳汇通量的37.5%。

如果取中国碳酸盐岩溶解消耗大气/土壤CO_2的估算通量为0.016 PgC/a；中国在1981–2000年，森林产生的净碳汇通量0.075 PgC/a，土壤碳汇通量0.04–0.07 PgC/a；则中国岩溶碳汇通量分别占森林碳汇通量的21.3%、土壤碳汇通量的22.9%–40%。

	Karst distribution area /($10^6 km^2$) 岩溶分布面积/($10^6 km^2$)	Karst carbon sink flux /(PgC/a) 岩溶碳汇通量/(PgC/a)	Carbon sink flux in terrestrial forest/(PgC/a) 陆地森林碳汇通量/(PgC/a)	Carbon sink flux in soil /(PgC/a) 土壤碳汇通量/(PgC/a)	Scale (karst/forest, karst/soil) /% 比例（岩溶/森林、岩溶/土壤）/%
全球	12.34–21.09	0.1–0.6（0.3）	1.7 ± 0.5	0.4–1.2	17.65，37.5
中国	3.44	0.016	0.075	0.04–0.07	21.3，22.9–40

Comparison of carbon sink flux in karst and in terrestrial ecosystem
岩溶碳汇通量与陆地生态碳汇通量间的对比

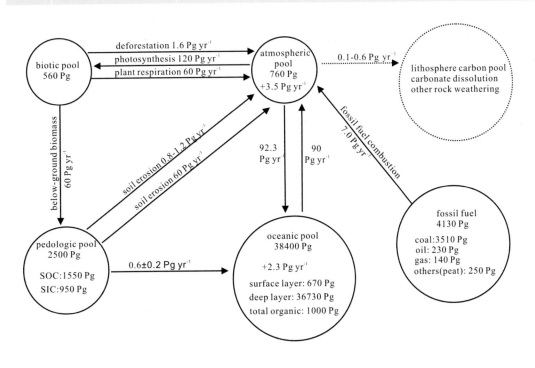

In the global carbon cycle model, the karst carbon sink has the same order of magnitude with that of terrestrial ecosystems
在全球碳循环模式中，岩溶碳汇的数量级可以与其他生态净通量处于同一数量级

Record of Paleoclimatic and Environmental Changes by Cave Stalagmite
岩溶洞穴石笋对过去气候环境变化的记录

The relationship between the signals of modern drip water and $\delta^{18}O$ value in stalagmite in Guilin Panlong Cave
桂林盘龙洞现代洞穴滴水与石笋 $\delta^{18}O$ 记录信号之间的关系

In 2009–2012, four perennial drip water sites in Panlong Cave, Guilin were monitored, with the monitoring indicators mainly including surface precipitation, cave drip water and $\delta^{18}O$ in modern speleothems (stalagmites). The results showed that in various places of the same cave, if the drip type and environment have the same conditions, the signals of $\delta^{18}O$ formed in carbonate ($CaCO_3$) are the same. They can validate each other. Signals recorded by modern carbonate sediments ($CaCO_3$) – stalagmite $\delta^{18}O$– mainly reflect precipitation, monsoon intensity.

2009—2012 年，在桂林盘龙洞中对 4 个常年性的滴水点进行了监测，主要监测的指标包括由地表降水、洞穴滴水到现代碳酸盐沉积物（石笋）中的 $\delta^{18}O$。检查结果显示，同一洞穴中不同地方在其滴水类型和环境相同条件下，所形成的碳酸盐（$CaCO_3$）的 $\delta^{18}O$ 的信号也是相同，具有平行相互验证的作用；现代碳酸盐沉积物（$CaCO_3$）——石笋 $\delta^{18}O$ 记录的信号主要反映降水及季风强度等气候信息。

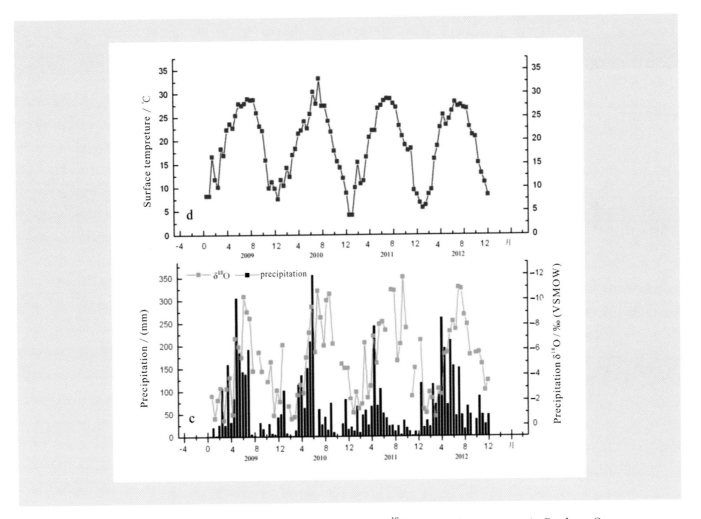

Variations of surface temperature, precipitation and $\delta^{18}O$ values of rain water, in Panlong Cave
盘龙洞地表气温、降雨及雨水中的 $\delta^{18}O$ 值变化

$\delta^{13}C$ value of modern carbonate sediments ($CaCO_3$) in Guilin Panlong Cave mainly reflecting the changes of surface ecosystem, as well as the impact of atmospheric precipitation and temperature
桂林盘龙洞现代碳酸盐沉积物($CaCO_3$)——石笋 的 $\delta^{13}C$ 值主要反映地表植被的变化，同时叠加了大气降水和温度的影响

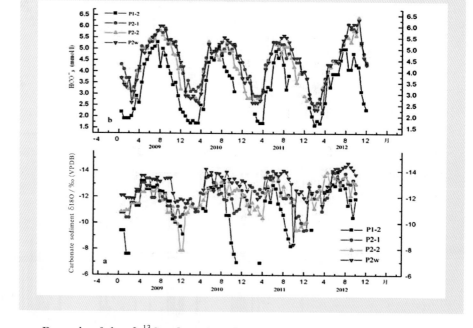

Variation of HCO_3^- concentration on of drip water, and $\delta^{13}C$ value of modern carbonate sediments in Panlong Cave

桂林盘龙洞洞穴常年性滴水的 HCO_3^- 与现代碳酸盐沉积物 $\delta^{13}C$ 值对比曲线图

Records of the $\delta^{13}C$ value of modern carbonate in the four perennial drip water sites showed that the $\delta^{13}C$ are lighter from late spring to early autumn, which is due to increasing at temperature precipitation and surface ecosystem activities change. Spring is the season when vegetation grows the fastest and has the largest amount of growth. In summer, vegetation will grow slowly. The growth of vegetation gradually declines due to dry weather in autumn and stops in winter.

4个常年性滴水点的现代碳酸钙(盐)沉积物的 $\delta^{13}C$ 值记录结果表明，在晚春至早秋季节，其 $\delta^{13}C$ 值偏负，这与降水增多、地表的生物量变化密切相关。春季是植被(物)生长速率最快，也是生物生长量最大的季节；其次是在夏季，植被(物)生长量不如春季；秋季后因气候干旱，植物的生长量逐渐下降；冬季植物的生长处于停滞状态。

High-resolution research on stalagmite revealed that during 100–150 ka BP, there existed an abrupt change process from glacial to interglacial periods, while show change from interglacial to glacial periods

高分辨率岩溶洞穴石笋的研究，揭示 10 万 –15 万年间，由冰期向间冰期的转化，是一个快速突变过程，由间冰期向冰期转化则是一个较缓的过程

Four stalagmites (D3, D4, D10 and D11) aged between 10ka BP to 15ka BP in Dongge Cave, Guizhou were collected, dating and their oxygen isotopes were measured. The results revealed that the era for the transition from glacial to interglacial periods at 6a/5e occurred in 129.28±0.45 ka BP. A rapid warming process was involved in the transition period in less than 200 years. The stalagmite $\delta^{18}O$ record suddenly became lighter from –5.5‰ to –8‰, with a change of 2.5‰, while temperature rose from 7.5 ℃ to 17.3 ℃, with a change of 9.7 degrees. During the transition process from the interglacial stage (5e) to glacial (5d) the oxygen isotope ratios changed from –7.34‰ to –4.91‰, with a change of 2.3‰, while the temperature dropped from 17.5 ℃ to 7.5 ℃, and even to the lowest level of 5 ℃. It took 2,700 years to complete the transition, thus showing a slower change process.

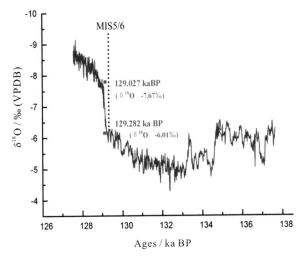

Low figure showed the rapid transition process from glacial to interglacial

由冰期 5e 向间冰期 5a 转化的较快过程

通过贵州荔波董哥洞 D3、D4、D10 和 D11 等 4 根石笋在 10 万 –15 万年，加密取样、定年和氧同位素测定，揭示 6a / 5e 时段冰期向间冰期转换年代为 129.28±0.45 ka BP，转换过程是一个快速回暖的过程，表现在小于 200 年的时间内，石笋 $\delta^{18}O$ 记录突然从 –5.5‰ 降低至 –8‰，变幅达 2.5‰，气温则从 7.5 ℃ 增至 17.3 ℃，变幅达 9.7 ℃；而由末次间冰期的间冰阶 (5e)/ 冰阶 (5d) 之间的转换，其氧同位素比值从 –7.34 ‰ 突变到 –4.91‰，变幅为 2.3‰，温度则由 17.5 ℃ 快速突变至 7.5 ℃，最低达 5 ℃，其转换年代用了 2700 年，表现为较为缓慢的转换特征。

Stalagmites in Dongge Cave recording the climate change pattern in Holocene from high temperature through adaptation to cooling down
贵州董哥洞石笋记录了全新世高温期、适应期和波动降温期的气候变化过程

According to ^{230}TH–U series age and oxygen and carbon isotope records from stalagmite D38, D4 in Dongge Cave, and Q6 in Qixing Cave, the climate change since 11,350 a BP can be divided into three stages: (1) early Holocene (11,350–8000 a BP) –high temperature period; (2) middle Holocene (8000–4500 a BP) –climate optimum period; (3) late Holocene (4500 a BP–present) –cooling period with fluctuations.

(1) Early Holocene–high temperature period. Stalagmite δ^{18}O varied from −9.43 to −7.37‰ (VPDB), with a 2.06‰ range and averaging −8.795‰ (VPDB). This is 0.494‰ lighter than −8.301‰. The δ^{18}O value for Whole of the Holocene Stalagmites(WHS), 1.015‰ lighter than −7.78‰ for modern cave drip water, and 1.265‰ lighter than −7.53‰ for modern carbonate (MC). Air temperature was 2 to 3℃ higher. In this period, stalagmite δ^{13}C varied from −8.74 to −2.73‰ (VPDB), with a range of 6.0‰, and averaging −4.72‰. This is 2.2‰ heavier than the WHS average δ^{13}C value of −6.91‰, and 3.9‰ heavier than −8.6‰ for MC.

(2) Mid–Holocene–Climate Optimum period. Stalagmite δ^{18}O varied from −9.28 to 7.77‰ (VPDB) in peak–valley–peak–valley fluctuations, with a range of 1.51‰ and averaging −8.665‰ (VPDB). This is 0.364‰ lighter than WHS, 0.885‰ lighter than modern cave drip water and 1.13‰ lighter than MC. Stalagmite δ^{13}C varied from −9.01 to −3.9‰ (VPDB), with a range of 5.1‰ and averaging −6.88‰ (VPDB). This is close to the WHS average δ^{13}C value, and 1.78‰ heavier than the average δ^{13}C value for modern sedimentary carbonates, but 2‰ lighter than in the early Holocene high temperature period, indicating that the ecological environment has improved. In this stage the East Asian summer monsoon was relatively strong, effective precipitation relatively increased, promoting the growth of woody vegetation, dominated by C3 plants (C3 plants accounted for more than 75%), with a warm and humid climate.

(3) Late Holocene–cooling with fluctuations period. The phase can be divided into seven sub-periods; three warm climate periods, three cold periods and a warming period. According to the stalagmites' ^{230}TH–U age versus oxygen isotope curves, the monsoon climate since 4500a BP can be roughly divided into seven climate (sub-) periods:

1. Warm period (4500–3000 a BP): Three temporary weak monsoon events recorded by stalagmite δ^{18}O were equivalent to the three temporary small cold events

recorded by stalagmite δ ^{13}C in northern Iberian Peninsula, Spain(4000,3550 and 3250 years) (Chivelet et al.,2011)

2. Iron Age Cold Period (3000–2750 a BP): The duration was 250 years in the Iron Cold Period in Europe, the same as Cold Period of Western Zhou Dynasty in China.

3. Moderate warm period (Roman Warm Period) (2750–2200 a BP): The duration was 550 years in Roman Warm Period in AD (Lamb,1997).

4. Cold interval (Dark Age Cold Period) (2200–1350 a BP): The duration was 850 years.

5. Warm period (Medieval Warm Period) (1350–700 a BP): The duration was 700 years in the Medieval Warm Period.

6. Cold period (700–150 a BP): The duration was 500 years in Europe (Lamb, 1997; Fagan, 2000), and the rest of the world, even in southern Africa or South America.

7. Modern warming period (150 a BP to present): Over the last 150 years, although Modern Climate Warming is faster than any transition periods, the warming was characterized by fluctuations rather than linear change.

根据贵州荔波董哥洞 D38、D4 石笋和都匀七星洞 Q6 石笋的 230Th-U 系年龄和氧、碳同位素记录将 11350 a BP 以来的气候变化，划分为 3 个气候变化阶段（右图）：①全新世早期（11350–8000 a BP）——高温期；②全新世中期（8000–4500 a BP）——气候适宜期；③全新世晚期（4500 a BP 以来）——气温波动的降温期：

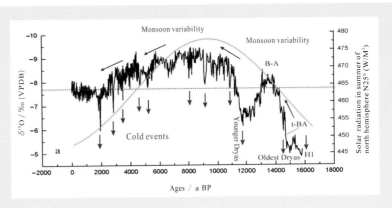
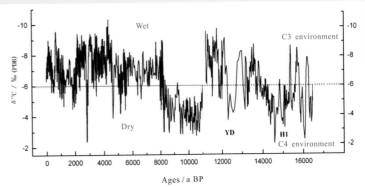
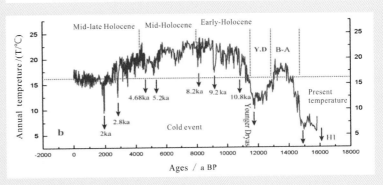

The Holocene climate change pattern of East Asian monsoon recording in stalagmites from Dongge Cave, Guizhou, China
从贵州董哥洞石笋中重建的全新世东亚季风气候变化模式

（1）全新世早期——高温期。石笋 $\delta^{18}O$ 变化在 –9.43‰至 –7.37‰（VPDB），变幅为 2.06‰，平均值（381 件）为 –8.795‰（VPDB），与全新世整段石笋的平均 $\delta^{18}O$ 值（–8.301‰）、现代洞穴滴水（–7.78‰）和现代沉积碳酸盐平均的 $\delta^{18}O$ 值（–7.53‰）分别要偏负 0.494‰、0.52‰和 0.77‰，气温要偏高 2–3℃；而本阶段石笋的 $\delta^{13}C$ 值变化在 –8.74‰至 –2.73‰（VPDB），变幅为 6.0‰，平均值为 –4.72‰（VPDB），与全新世整段石笋的平均 $\delta^{13}C$ 值（–6.91‰）及现代沉积碳酸盐的平均值（8.6‰）相比，分别偏重或正 2.2‰和 3.9‰。本阶段反映东亚夏季风相对强盛，气温升高，有效降水相对较少，不利于木本植被生长，以 C4 植物为主（占 60%），气候以干热气候环境为特征。

（2）全新世中期——气候适宜期，石笋 $\delta^{18}O$ 变化在 –9.28‰至 –7.77‰（VPDB），呈峰–谷–峰–谷的波动变化，其变幅为 1.51‰，平均值（152 件）为 –8.665‰（VPDB），与全新世整段石笋的平均值（–8.301‰）、现代洞穴滴水（–7.78‰）和现代沉积碳酸盐平均值（–7.53‰）相比，分别偏负或偏低 0.335‰、0.885‰和 1.13‰；而石笋 $\delta^{13}C$ 变化在 –9.01‰至 –3.9‰（VPDB），其变幅为 5.1‰，平均值为 –6.88‰（VPDB），与全新世整段石笋的平均 $\delta^{13}C$ 值（–6.91‰）及现代沉积碳酸盐的平均值（8.6‰）相比，本阶段石笋的 $\delta^{13}C$ 值接近全新世整段石笋的平均 $\delta^{13}C$ 值，比现代沉积碳酸盐的平均 $\delta^{13}C$ 值稍偏重 1.78‰，但是比全新世早期——高温期偏轻或负 2‰，其生态环境有所好转。本阶段反映东亚夏季风相对强盛，有效降水相对增大，利于木本植被生长，以 C3 植物为主（C3 植物占 75%以上），为温暖湿润气候环境。

（3）全新世晚期——气温波动的降温期。可分为 7 个气候亚期：3 个温暖期、3 个寒冷期和 1 个升温期。

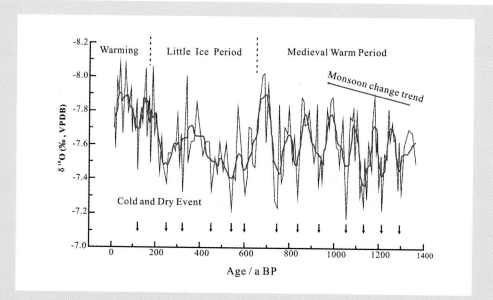

The climate change pattern over the last 1400 years of East Asian monsoon recording in stalagmites from Dongge Cave, Guizhou, China
从贵州董哥洞石笋中重建的 1400 年以来东亚季风气候变化模式

According to the curves of stalagmites' 230 Th-U ages and oxygen isotope, the monsoon climate since 4500 a BP can be roughly divided into seven climatic (sub-) periods : ① warm period between 4500 a BP and 3000 a BP; ② Iron Age Cold Period from 3000–2750 a BP; ③ moderate warm period (Roman Warm Period from 2750–2200 a BP; ④ cold interval (Dark Age Cold Period) period from 2200 –1350 a BP; ⑤ warm period (Medieval Warm Period) from 1350–700 a BP ; ⑥ cold period from 700–150 a BP; ⑦ Modern Warming period from 150 a BP until now.

1. Three temporary weak monsoon events recorded by stalagmite $\delta^{18}O$ were equivalent to the three temporary small cold events recorded by stalagmite $\delta^{13}C$ in northern Iberian Peninsula, Spain (4000, 3550 and 3250 years) (Chivelet et al., 2011);

2. The duration was 250 years in the Iron Cold Period in Europe, the same as Cold Period of Western Zhou Dynasty in China;

3. The duration was 550 years in Roman Warm Period in AD (Lamb, 1997);

4. The duration was 850 years;

5. The duration was 700 years in Medieval Warm Period;

6. The duration was 500 years in Europe(Lamb, 1977; Fagan, 2000), and the rest of the world, even in southern Africa or South America;

7. Over the last 150 years, although Modern Climate Warming is faster than any transition periods, the warming was characterized by fluctuation rather than linear change.

根据石笋的230Th-U系年龄和氧同位素曲线的波动形式，可将4500 a BP以来的季风气候变化，大致可分为 7 个气候 (亚) 期：① 4500–3000 a BP 温暖期；② 3000–2750 a BP 寒冷期 (Iron Age Cold Period)；③ 2750–2200 a BP 现代温暖期；④ 2200–1350 a BP 寒冷期；⑤ 1350–700 a BP 温暖期；⑥ 700 –150 a BP 寒冷期；⑦ 150 a BP 至今，升温期 (Modern Warming)。

1. 石笋 $\delta^{18}O$ 记录的 3 个短暂的弱季风事件与西班牙伊比利亚北部洞穴石笋的 $\delta^{13}C$ 记录的 3 次短暂的小冷事件 (4000 年、3550 年和 3250 年) 相当 (Chivelet et al., 2011)；

2. 持续时间为 250 年欧洲的铁器冷时代在中国历史文献中记载的西周寒冷期；

3. 持续时间为 550 年相对于著名的罗马暖期 (Roman Warm Period)(Lamb, 1997)；

4. 持续时间为 850 年；

5. 持续时间为 700 年，应相对于中世纪温暖期 (Medieval Warm Period)；

6. 持续时间为 500 年，全欧洲均有广泛报道 (Lamb, 1977; Fagan, 2000)，并在世界其他地区，甚至于南非或南美也可能存在；

7. 在最近 150 年，虽然"现代气候变暖"比以往任何过渡期都快，但是变暖并不是呈线性的，而是呈锯齿状波动。

Water Cycle: Karst Dynamic System, Water Cycle and Water Resources
水循环：岩溶动力系统与水循环、水资源

Water provides the essential resources for life on Earth surface systems. It has close relationship with the formation and evolution of atmospheric system, human society and natural environment. More importantly, fresh water is a vital resource for human survival, health, social prosperity and security, and maintains the sustainable development of human society. Therefore, the Natural Sciences Sector at UNESCO set up the International Hydrological Programme (IHP) to carry out water research, water resources management and water-related educational activities. Karst water belongs to groundwater.

The Chinese government has been focused on the survey, research, exploitation, utilization and management of karst water. More than 3,000 underground rivers have been distributed in karst areas in southwest China, with a total length of over 14,000 km, and recharge area of approximately 300,000 km^2. There are even 47 billion m^3/a of runoff in dry season, which is equivalent to the runoff in the Yellow River.

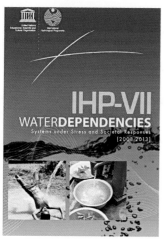

水是地球生命系统的源泉，水关系着地球大气系统、人类社会和自然环境的形成、演化。淡水更是关系着人类生存、健康，社会繁荣和安全的关键资源，维系着人类社会的可持续发展。为此，在联合国教科文组织自然科学部专门设立了国际水文计划（International Hydrological Programme，IHP），开展水科学研究、水资源管理及与水相关的教育活动。岩溶水属于地下水。

中国政府一直关注岩溶水的调查、研究，开发利用和管理。经调查中国西南岩溶区有3000多条地下河，总长度超过1.4万km，汇水面积约达30万km^2，枯水季节径流量也高达470亿m^3/a，相当于一条黄河的径流量。

Scientific research projects related to water cycle:

(1) Land and Resources Survey Work Project "Key environmental and geological issues and response options" (1212010813111), 2008-2010, Yuan Daoxian;

(2) Land and Resources Survey Work Project "Integration of environmental and geological survey and maps development for karst mountainous areas in southwest China" (1212010813112), 2008-2010, Shi Jian;

(3) Land and Resources Survey Work Project "Underground water exploration in the areas in Shandong that are severely short of water" (1212011121178), 2010, Tang Jiansheng, Li Zhaolin;

(4) Land and Resources Survey Work Project "Hydro-geological and environmental survey of typical karst river basins in southwest China" (1212011220950), 2012-2013, Tang Jiansheng, Li Zhaolin;

(5) Land and Resources Survey Work Project "Survey and dynamic research of typical subterranean streams in southwest China" (1212011220959), 2012-2015, Yi Lianxing;

(6) Land and Resources Survey Work Project " Integrated hydro-geological and environmental research and information system development for karst areas in southwest China" (1212011121157), 2011-2015, Xia Riyuan;

(7) Land and Resources Survey Work Project "Karst geological data integration and information service system development" (1212011220355), 2012-2015, Shi Jian.

近年来与水循环相关的科研项目：

[1] 中国地质调查局项目"西南岩溶石山地区重大环境地质问题及对策研究"（1212010813111），2008-2010，袁道先；

[2] 中国地质调查局项目"西南岩溶石山地区环境地质调查综合集成和图系编制"（1212010813112），2008-2010，时坚；

[3] 中国地质调查局项目"山东严重缺水地区地下水勘查"（1212011121178），2010，唐建生、李兆林；

[4] 中国地质调查局项目"西南典型岩溶流域水文地质及环境地质调查（云南文山广南4幅）"（1212011220950），2012-2013，唐建生、李兆林；

[5] 中国地质调查局项目"西南典型岩溶地下河调查与动态研究"（1212011220959），2012-2015，易连兴；

[6] 中国地质调查局项目"西南岩溶地区水文地质环境地质综合研究与信息系统建设"（1212011121157），2011-2015，夏日元；

[7] 中国地质调查局项目"岩溶地质数据集成与服务系统建设"（1212011220355），2012-2015，时坚。

Water cycle in karst areas characterized by vertical infiltration, and underground water resources
岩溶区水循环以垂向下渗为特征、水资源以地下水资源为主

Statistical data showed that: the density of surface river system in karst area is 0.23 km/km^2, 0.35 km/km^2 in non-karst area; the number of underground rivers with length over 2 km and the density 0.1 km/km^2 is 435. The average volume of water resources in karst region and non-karst region is 720,000 and 950,000 m^3/km^2. The average volume of groundwater resources in karst region and non-karst region is 500,000 m^3/km^2 and 220,000.35 m^3/km^2.

统计结果显示：岩溶区地表河的密度0.23 km/km^2，非岩溶区0.35 km/km^2；长度>2 km的地下河435条，密度0.1 km/km^2。岩溶区平均水资源量为720 000 m^3/km^2，非岩溶区950 000 m^3/km^2；岩溶区平均地下水资源量为500 000 m^3/km^2，非岩溶区220 000.35 m^3/km^2。

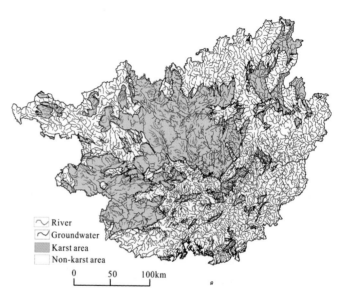

Distribution of surface river system and underground river system in Guangxi
广西岩溶区、非岩溶区地表河与地下河的空间分布图

With the water circulation pattern, the hydrological systems in karst areas are significantly different from those in clastic zones.

Karst hydrological systems are characterized as ground river system. Surface precipitation that dominated vertically moves into saturation zone and subterranean streams from the epikarst zone and vadose zone. Most of the precipitation directly penetrates into subterranean streams through, sinkholes, shafts into ground conduit, then it joins the surface rivers at the lower areas. In this way, the large watersheds in karst areas are rich in surface water resources, while the small watersheds, especially underground river basins are short of surface water resources; on the other word,

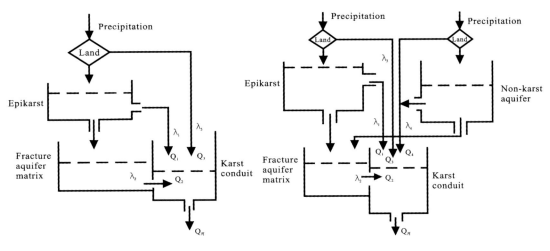

(a) Shetch of water cycle in pure karst aquifer (b) Sketch of water cycle in pure karst aquifer coupling with non-karst area

the ground water resources are separated by the surface soil resources. As for the spatial distribution of water resources, the karst areas are lack of surface water systems, while underground water systems are well developed.

The hydrological systems in clastic zones are mainly composed of surface water system. Precipitation, through overland flow, forms surface water systems, which makes it easier to conduct artificial regulation as well as storage and efficient use.

从水循环模式方面来看，岩溶区水文系统与碎屑岩区的水文系统具有如下显著差异。

岩溶水文系统以地下水为主，降水通过岩溶表层带、包气带垂向运移至饱水带和地下河。其中很大部分降水则直接通过漏斗、落水洞、竖井，以管道流的方式直接进入地下河，然后通过地下河在较远的低洼处再汇入地表河，这就出现岩溶区大流域水资源丰富，而小流域，尤其是地下河流域内水资源短缺、水土资源不协调的尴尬状况。从水资源的空间分布情况来看，岩溶区十分缺乏地表水系，但地下水系发育。

而碎屑岩区的水文系统则以地表水为主，降水通过坡面产流，形成纵横交错的地表水系网络，这给水资源的人工调蓄和有效利用带来便利。

Success case of exploitation and utilization of epikarst springs
岩溶表层带泉的开发利用的成功案例

In Zerong Town, Xingyi County, Guizhou Province, 3532 small ponds and 2630 small water tanks have been built with epikarst springs and overland flow. The water tanks in upstream and downstream are linked with each other to improve the water regulation and storage capabilities. To save land resources, many tanks are directly built on sloping lands, using roads and furrows flows to store water. Some tanks are even built in the rice paddy fields to store rainwater and excess water of paddy fields. In addition, some cellars are built in front of and behind the houses to store rainwater for drinking. To make full use of land resources, the local people have converted tops of water tanks into paddy fields, stony grounds into flat paddy fields, and slopes into terraced fields. Through the production of food crops, the cultivation of economic plants to solve the drinking water and some irrigation water, the income of local farmers is increased. These efforts have raised the food production, cultivated economic crops, solved the difficulties of irrigation and drinking water, and increase the local farmers' incomes.

贵州省兴义县则戎乡利用岩溶表层泉、山坡坡面流，建起了大大小小的水窖和水池，修建了3532个小水池、2630个小水窖；为了提高水资源的调蓄能力，上游的水窖与下游的水窖相连；为了节约土地资源，很多水窖就直接建在坡耕地间，并利用路面和汇流沟集水于水窖中，甚至将水窖建在稻田中贮集稻田里的雨水和多余的稻田水，屋前屋后建水窖，集雨水作饮用水源；为了充分利用土地资源，将水窖顶改造成水田，石旮旯地改造成平整的水田，坡地改变为梯田。通过粮食作物的增产、经济植物的栽培，解决了人畜饮水和部分灌溉用水，增加了农民的收入。

Case on exploitation and utilization of subterranean stream water resources
地下河水资源的开发利用典型案例

Statistical data of karst groundwater resources in provinces in southwest China
我国西南省（区）岩溶地下水资源统计表

Province 省（市区）	Natural water resources /($\times 10^8 m^3/a$) 多年平均天然资源量 /($\times 10^8 m^3/a$)	Allowable amount /($\times 10^8 m^3/a$) 允许开采量 /($\times 10^8 m^3/a$)	Exploited amount /($\times 10^8 m^3/a$) 已开采量 /($\times 10^8 m^3/a$)	Percentage of exploited/% 已开采比例/%
Yunan 云南	215.7	57.39	2.16	3.77
Guizhou 贵州	380.2	138.87	16.03	11.54
Hunan 湖南	250.92	69.09	9.42	13.64
Guangxi 广西	555.64	166.15	13.59	8.18
Total 合计	1402.46	431.5	41.2	9.55

There are a large amount of groundwater resources of karst areas in southwest China. The total amount that has been exploited in four provinces in southwest China only accounts for 9.55% of allowable groundwater resources. Since karst areas are severely short of surface water, therefore, the supply of groundwater resources becomes extremely important. Rational and effective exploitation and utilization of underground water resources can not only make the surface land use in good patterns and agricultural structure, but also alter local energy structure, which can generate significant economic and social benefits. For example, use of the waters in aquaculture and in hydro-power generation improves renewable green energy resources.

中国西南岩溶区地下水资源总量大，但开发利用程度低，四省区已开采总量仅占允许开采量的9.55%。由于岩溶区地表水资源的严重缺乏，地下水资源的供水意义显得极为重要。合理、有效开发利用地下河水资源，不仅可以改变地表土地利用方式，转变农田产业结构，同时可以改变当地能源结构，具有显著的经济效益及社会效益，如利用水域养殖水产，利用水能发电，提高可再生的绿色能源的比例。

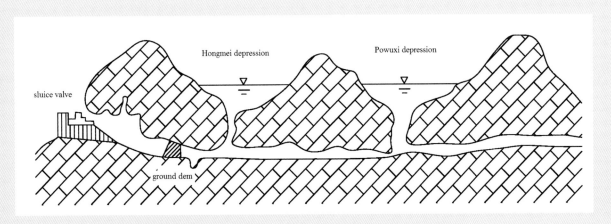

Sketch of Fenfa underground reservoir in Xinzhai Village, Dushan County, Guizhou
贵州独山新寨奋发洞地下水库示意图

When the ground dam was built, the water level of Fenfa underground reservoir in Xinzhai Village, Dushan County, Guizhou Province, rose by 26 m. It is linked with Hongmei Depression and Powuxi Depression that are connected with subterranean streams to form a lake. The underground reservoir has a storage capacity of 220,000 m^3, which can be used to irrigate a total area of 100 hm^2 of farmlands, thus solving the local water supply and irrigation problems.

贵州独山新寨奋发洞地下水库，其水位提高 26 m，在上游与地下河连通的红梅洼地和破屋西洼地相连蓄水成湖，地下库容达 22 万 m^3，引流自流灌溉 100 hm^2，解决了当地供水及灌溉问题。

Impact of surface land use pattern on water quantity and amount of subterranean streams
地表土地利用方式对地下河水质、水量的影响

Lingshui Spring, as a covered karstic spring with 697 km^2 of recharge area, provides drinking water source for Wuming County (currently about 200 000 people, which is expected to grow to 500 000 in the next decade). Over the past 30 years, although the annual precipitation maintained 1000–1500 mm, the water flow of spring discharge declined from 4000 L/s in 1977 to 2600 L/s in 2001, while large amount of waterweeds perished at the spring outlet.

Preliminary analysis indicated that this might be related to the change of land use pattern in the recharge area. Since late 1980s and early 1990s, in order to seek economic benefits, the natural vegetation in the recharge areas has been gradually replaced by eucalyptus forests, whose growth needs to consume large amount of groundwater. Monitoring data on three representative eucalyptus forest plots in Guangdong showed that the water amount consumed annually to grow eucalyptus was recorded as 443 mm, 402 mm and 363 mm respectively, representing 26.7%, 24.5% and 29.0% of local annual rainfall. Meanwhile, eucalyptus can produce a large amount of bio-chemical substances which may cause negative impacts on water environment.

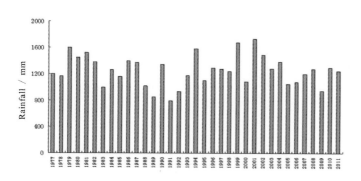

Annual precipitation from 1977 to 2011

1977 – 2011 年的年降水量

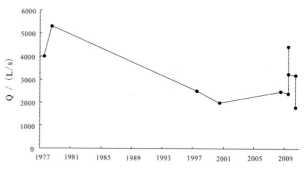

Runoff of Lingshui Spring in dry season

灵水泉枯季径流量

灵水泉是覆盖型的岩溶泉，补给面积 697 km^2，它是广西武鸣县县城 20 万人的饮用水水源（随着城镇化发展，该县城的人口在未来的 10 年将发展到 50 万）。在过去的 30 年间，虽然年降雨量没有很大的变化，约 1000–1500 mm，但该泉的枯季流量从 1977 年的 4000 L/s，降低到 2001 年的 2600 L/s，同时泉口的水草出现大量死亡的现象。

初步分析其原因，可能与补给区土地利用方式的改变相关，为了追求经济效益，在 20 世纪 80 年代末至 90 年代初，补给区的自然植被逐渐被桉树林取代，而桉树的生长则需要消耗大量的地下水。广东 3 个代表性桉树林的监测结果显示，桉树生长每年消耗的水量分别为 443 mm、402 mm 和 363 mm，分别占当地年降雨量的 26.7%、24.5% 和 29.0%。同时，桉树可分泌大量的生物化学物质，对水环境产生影响。

左图
Left
Outlet of Lingshui Spring.
灵水泉泉口。

Integration of karst hydrogeological survey and geophysical technologies playing an active role in water exploration in southwest China
岩溶水文地质的综合调查与地下水综合探测技术有机结合在中国西南地区应急抗旱找水中发挥积极作用

From the fall of 2009 to the summer of 2010, Southwest China suffered from severe drought, mainly in karst regions in Yunnan, Guizhou, and Guangxi. The amount of rainfall was reduced by half that in the same period of the previous year, and the lowest record for the past 50 years. Eventually, there were 5 million local people having no access to drinking water and 3.5-million-ton reduction of grain caused by the severe drought.

To find the emergency water against drought, the China Ministry of Land and Resources and China Geological Survey organized 10,000 experts, including 2600 geologists and hydrogeologists to seek groundwater source in the disaster-stricken areas.

The target water sites were defined according to 1:50,000 hydrogeologic, environmental geology surveys maps. The karst hydrogeological conditions were used to assess karst groundwater recharge and convergence conditions. In the target water site, a geophysical method-Horizontal Electrical Joint Profiling (HEP) was employed to detect horizontal position and width of underground hydrogeologic body, while high-density and symmetrical Vertical Electrical Sounding (VES) detection method was applied to probe the depth of groundwater. As a result, 2348 wells have been successfully drilled (including 1971 wells of karst underground water) with a total water amount of 360,000 m^3/day. Thus these helped solve the problems of providing drinking water for 5,200,000 people in the local areas.

2009年秋天至2010年夏，西南遭受持续干旱，其西南主要岩溶发布省云南、贵州、广西的降雨量比同期减少了一半，是近50年的最低纪录。受灾最为严重地区与岩溶区吻合。当地居民5 000 000人口缺乏饮用水，粮食减产3 500 000吨。

为了应急抗旱找水，国土资源部、中国地质调查局共同组织了10 000人，其中地质、水文地质专业人员2600人，到灾区找水。

根据1:50 000水文地质、环境地质调查成果，拟定找水靶区，根据岩溶水文地质条件，分析岩溶地下水补给和汇水条件；在靶区用地球物理方法——视电阻率联合剖面法 Horizontal Electrical Profiling (HEP) 探测地下地质体的水平位置和宽度；用高密度对称四极电测深探测法 Vertical Electrical Sounding (VES)，探测地下水的埋深。成功地钻孔打井找水2348处，总出水量达360 000 m³/d，解决5 200 000人的饮用水问题。其中有1971处为岩溶地下水。

Calcium Cycle: Basis for Karst Ecosystem
钙循环：岩溶生态系统研究的基础

As a unique element in karst ecosystem, calcium provides a lot of base cations in the soil for plants. According to the response to calcium, plants can be divided into calciphilous plants and calcifuge plants.

The calcium in limestone soil shows strong capacity of competing with other cations, which influences the activity of other metal cations, and restricts the absorption of these elements by plants.

Therefore, the study on the nutrient elements migration in karst ecosystem depends largely on the study of calcium cycle. Understanding the pattern of calcium cycle plays a crucial role in guiding the soil improvement and optimum selection of suitable plants in the process of rocky desertification control. Unfortunately, the study on the calcium cycle and its impacting factors is still very weak and needs to be further enhanced.

钙是岩溶生态系统中最具特色的元素，它是土壤中的

Higher plants directly grow on carbonate rocks. Their roots pass throug rock crevices, obtain nutrients from the calcium-rich carbonate rocks, and are affected by calcium-rich environment.

高等植物直接生长在碳酸盐岩上，根系穿插在岩石缝隙中，植物从富钙的碳酸盐岩中获得养分，同时也受到富钙环境的影响。

盐基离子、植物必需的大量营养元素，因此根据植物对土壤钙盐响应的关系，可将植物分为喜钙植物和嫌钙植物。

有理由相信，在石灰土环境中钙表现出较强的与其他阳离子竞争有机配位的能力，影响了其他金属离子的活性，并制约植物对这些元素的吸收。

因此，岩溶生态系统中元素迁移规律的研究，依赖对钙循环的研究，对钙循环规律的掌握将在石漠化综合治理过程土壤改良和适宜性植物优选具有重要的指导作用。遗憾的是钙的循环规律及对资源环境的影响的研究还比较薄弱，需要进一步加强。

The projects related to calcium cycle since 2008:

(1) Chinese National Natural Science Foundation of China (General Program) "Soil calcium migration in karst dynamic system and its eco-environmental benefits" (40872213), 2009.1–2011.12, Cao Jianhua；

(2) Chinese Natural Science Foundation of China Program for Young Scholars "Mechanism of calcium in regulating the photosynthetic physiology of *Lonicera japonica* in the karst drought process" (41003038), 2011.1–2013.12, Li Qiang.

与钙循环相关的科研项目：

[1] 国家自然科学基金面上项目 "岩溶动力系统中土壤钙迁移及生态环境效益" (40872213)，2009.1–2011.12，曹建华；

[2] 国家自然科学基金青年基金 "岩溶干旱过程中钙对忍冬光合生理的调控机制" (41003038)，2011.1–2013.12，李强。

Content of easily-migrated calcium accounting for 80% of limestone soil in Maocun karst experimental site, Guilin
桂林毛村岩溶地下河流域石灰土中易迁移的钙含量占 80%

The typical limestone soil and red soil sections in Maocun, Guilin were sampled in different layers with a depth of 1 m, and the improved six-step Tessier sequential extraction method was used to analyse the calcium species in soil. It was initially found that:

Total amount of calcium accounts for 0.7% to 0.9% in limestone soil, which is present mainly in the forms of ion exchange state (including water soluble state), carbonate combined state and strong organic combined state (including some sulfide state). The ion exchange state (including water soluble state) takes up a majority (87.1% of the total amount), which is followed by the carbonate combined state (8.47% of total amount) and strong organic combined state (including some sulfide state) (2.26% of total content). Contents of ion exchange state in different layers of limestone soil have little difference, indicating their strong migration. Exchangeable calcium is one of the major base cations in soil. It can be substituted by exchangeable ions in solutions, and maintain a dynamic balance with soluble Ca, highlighting its Ca-rich feature in karst areas. The combined occupation of Ca in the other three states account for less than 3% of the total content of Ca.

In red soil profile, Ca content takes up 0.003%–0.04%, being 1–2 orders of magnitude lower than that of the limestone soil. Most of soil calcium is concentrated at the top of red soil section, mainly in the form of iron and manganese oxides combined state. Among them, iron and manganese oxides, ion exchange state (including water soluble state), humic acid combined state, strong organic combined state (including some sulfide state), carbonate combined state and residual state take up 30.37%, 27.33%, 14.01%, 12.39%, 12.02% and 3.88% of the total amount, respectively.

桂林毛村典型石灰土和红壤的剖面上，按层位取样，深度 1 m，采用改进的 Tessier 六步提取法分析土壤钙赋存的形态，初步获得以下结果：

石灰土剖面中 Ca 元素的总量为 0.7%-0.9%，主要是以离子交换态（包括水溶态）、碳酸盐结合态和强有机结合态（包括部分硫化物态）存在，特别是以离子交换态（包括水溶态）的含量为主（占总含量 87.1%），其次以碳酸盐结合态（占总含量 8.47%）和强有机结合态（包括部分硫化物态）（占总含量 2.26%）存在。离子交换态在石灰土剖面各层中存在的含量相差不大，说明其迁移性很强，交换态 Ca 是吸附于土壤胶体表面的 Ca 离子，是土壤中主要的盐基离子之一；交换态 Ca 能被溶液中的交换性离子替换下来，因此和溶液态 Ca 保持着动态平衡，显著突出了岩溶区石山富 Ca 的特点；其余三个形态之和占总含量不到 3%。

红壤剖面中 Ca 元素的总量为 0.003%-0.04%，与石灰土相差 1-2 个数量级。大部分土壤钙集中在红壤剖面顶部，主要是以铁锰氧化物结合态形式存在。其中，铁锰氧化物结合态、离子交换态(包括水溶态)、腐殖酸结合态、强有机结合态（包括部分硫化物态）、碳酸盐结合态和残渣态分别占总量的 30.37%，27.33%，14.01%，12.39%，12.02% 和 3.88%。

Calcium species distribution in limestone soil profile
石灰土土壤剖面中钙形态分布特征

Depth/cm 剖面深度/cm	Ca Amount/(mg/kg) Ca 含量/(mg/kg)						
	A	B	C	D	E	F	T
0–15	7575	1115.5	109.16	92.9	219.6	15.39	9127.55
15–30	7500	755.5	76.04	90.3	82.14	19.55	8523.53
30–45	6675	692.5	70.44	77.7	241.6	21.64	7778.88
45–60	8110	598.5	65.92	31	81.48	22.16	8909.06
60–75	6525	515	63.12	78.1	277.2	22.67	7481.09
75–90	6305	537	64.44	112.3	84.46	21.97	7125.17

Note: A – Ion exchangeable state (including water soluble state), B – carbonate-combined state, C – humic acid combined state (loosely organic combined state), D – iron-manganese oxides, E – strong organic combined state(including some sulfide state), F–residual state, T–sum of the content of different states.

注：A 为离子交换态（包括水溶态），B 为碳酸盐结合态，C 为腐殖酸结合态（松结有机结合态），D 为铁锰氧化物结合态，E 为强有机结合态（包括部分硫化物态），F 为残渣态，T 为各形态含量之和。

Calcium species distribution characteristics in red soil profile
红壤土壤剖面中钙形态的分布特征

Depth/cm 剖面深度/cm	Ca Amount/(mg/kg) Ca 含量/(mg/kg)						
	A	B	C	D	E	F	T
0–15	163.5	54.28	67.4	98.4	48.96	9.39	441.93
15–30	1.52	3.92	2.65	23.33	27.62	2.18	61.22
30–45	3.64	5.55	6.61	31.7	5.54	2.14	55.18
45–60	2.72	1.9	3.48	18.82	3.1	1.94	31.96
60–75	4.61	3.58	5.39	21.43	1.91	2.11	39.03
75–90	16.5	9.68	10.49	15.12	2.11	3.24	57.14

Note: A – Ion exchangeable state (including water soluble state), B – carbonate-combined state, C – humic acid combined state (loosely organic combined state), D – iron-manganese oxides, E – strong organic combined state(including some sulfide state), F–residual state, T–sum of the content of different states.

注：A 为离子交换态（包括水溶态），B 为碳酸盐结合态，C 为腐殖酸结合态（松结有机结合态），D 为铁锰氧化物结合态，E 为强有机结合态（包括部分硫化物态），F 为残渣态，T 为各形态含量之和。

Calcium content in limestone soil affecting the pH of soil and content of organic soil carbon
石灰土中钙含量影响着土壤的 pH 和有机碳的含量

Forty four samples were collected from different topographic positions (i.e., depression, hillside, pass and plain) in spring, summer, autumn and winter in Yaji, Guilin, by which the soil calcium content, pH of soil, and content of soil organic carbon were tested. The results showed that the total amount of calcium in descending order is black limestone soil > brown soil> yellow limestone soil. Content of soil organic matter and total amount of calcium in soil are significantly positively correlated, with a correlation coefficient of 0.744.

Top right figure showed there is a linear relationship between the total calcium content in soil and pH of soil, with the correlation coefficient 0.771. They are significantly positively correlated ($P<0.01$). It suggests the total calcium content and pH grow in a coordinated way.

对桂林丫吉不同地貌部位，洼地、山坡、垭口、平原按照不同季节采集 44 个样品，对土壤钙含量、土壤 pH 及有机碳含量进行检测，检查结果显示钙总量由高到低的顺序为黑色石灰土 > 棕色石灰土 > 黄色石灰土；土壤有机质含量与土壤总钙含量存在显著正相关关系，相关系数为 0.744。

由右上图可知，土壤总钙含量与土壤 pH 含量存在线性关系，利用 SPSS 17.0 软件计算桂林丫吉试验场土壤总钙含量与土壤 pH 的 Pearson 相关系数 ($n=44$)，结果表明：相关系数为 0.771，数据相关性为显著正相关水平 ($P<0.01$)，说明总钙与 pH 是协同发展的。

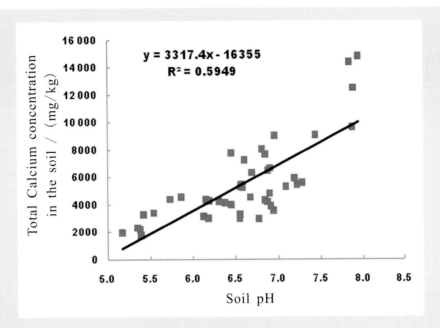

Linear relationship between soil total calcium and soil pH in Yaji Karst Experimental Site, Guilin

桂林丫吉岩溶试验场土壤总钙与pH线性关系

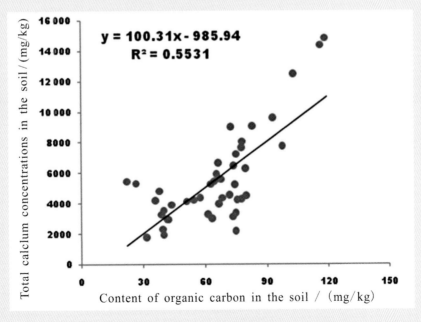

Linear relationship between soil total calcium and soil organic matter content in Yaji Karst Experimental Site, Guilin

桂林丫吉岩溶试验场土壤总钙与有机质含量线性关系

Calcium species in calciphilous plant leaves in karst areas is dominated by calcium pectate and mainly occuring in cell walls
岩溶区嗜钙型植物叶片中钙以果胶酸钙形态为主，且主要赋存在细胞壁

Banzhai underground watershed in karst areas (limestone, dolomite) of Guizhou Maolan National Natural Reserve and Yaopai watershed in non-karst areas (sandstone, shale zone) were selected to conduct a comparative study on the species of calcium in plant leaves under different geological conditions. Thirteen different plant species each was collected from the karst and non-karst areas. There were six endemic species from karst and non-karst areas, respectively. The total contents of calcium in leaves, their forms and distribution (subcellular components) were tested. The results showed that: ① the average content of calcium in the plant leaves from the karst areas was 1216.82 mg/kg, 58.45% higher than that from the non-karst areas; ② pectate calcium, which accounts for 27.91%, 32.82% of the total calcium in the leaves of calciphile plants, is the main form of calcium in the calciphile plant leaves in the karst areas; whereas calcium oxalate, which accounts for 33.69%, 34.34% of the total calcium in the calcifuges plant leaves, is the main form of calcium in the calcifuge plant leaves in the non-karst areas; ③ 59.05%–66.54% of the total calcium exists in cell-wall of calciphile plant leaves in the karst area, while 36.67%–43.77% of the total calcium exists in cytoplasm of the calcifuge plant leaves in non-karst areas.

选择贵州茂兰国家级自然保护区岩溶区（石灰岩、白云岩）板寨地下河流域、非岩溶区（砂岩、页岩区）尧排流域，对不同地质条件下植物叶片中钙形态进行对比研究；采集岩溶区、非岩溶区植物品种各13种，其中岩溶区、非岩溶区特有植物品种各6种，分析其叶片中钙质总量、形态及分布部位（亚细胞组分）。结果显示：①岩溶区植物叶片钙质含量平均为1216.82 mg/kg，比非岩溶区高出58.45%；②岩溶区嗜钙型植物叶片中钙以果胶酸钙形态为主，其含量占总钙质量的27.91%–32.82%；而非岩溶区嫌钙型植物叶片中的钙质以草酸钙形态为主，占总钙质量的33.69%–34.34%；③岩溶区嗜钙型植物叶片中的钙主要赋存在细胞壁中，占总钙质的59.05%–66.54%，而非岩溶区嫌钙型植物叶片中的钙主要赋存在胞质中，占总钙质的36.67%–43.77%。

Mean calcium content and percentage of different calcium species in total amount of calcium of leaves (mg/kg)
不同类型植物叶片钙形态含量均值及百分比（mg/kg）

Type 分组		Calcium nitrate and Calciium oxide 硝酸钙和氧化钙	Soluble calcium 水溶性钙	Pectin 果胶酸钙	Calcium phosphate and Calsium carbonate 磷酸钙和碳酸钙	Calcium oxalate 草酸钙	Calcium silicate 硅酸钙	Total 总量
Calciphile 钙型	Old leaf 老叶	35.3 (2.45)	238.5 (16.53)	473.64 (32.82)	154.69 (10.72)	338.59 (23.46)	202.35 (14.02)	1443.07
	Young leaf 嫩叶	21.44 (1.58)	216.54 (15.93)	379.55 (27.91)	206.28 (15.17)	325.8 (23.96)	210.08 (15.45)	1359.69
Moderate (Karst Region) 中间型（岩溶区）	Old leaf 老叶	29.23 (2.77)	53.63 (5.09)	102.46 (9.27)	103.34 (9.81)	264.76 (25.13)	500.31 (47.48)	1053.73
	Young leaf 嫩叶	33.2 (3.05)	34.5 (3.17)	41.21 (3.78)	72.46 (6.65)	245.08 (22.49)	663.43 (60.87)	1089.88
Moderate (non-Karst Region) 中间型（非岩溶区）	Old leaf 老叶	24.68 (2.95)	50.83 (6.07)	88.77 (10.6)	68.48 (8.81)	201.16 (24.02)	403.39 (48.18)	837.31
	Young leaf 嫩叶	19.58 (2.28)	28.23 (3.29)	61.85 (7.21)	48.39 (5.64)	225.1 (26.23)	475.02 (55.35)	858.17
Caleifuge 嫌钙型	Old leaf 老叶	19.98 (2.74)	135.29 (18.55)	105.7 (14.49)	81.46 (11.17)	245.69 (33.69)	141.12 (19.35)	729.24
	Young leaf 嫩叶	13.43 (2.13)	50.81 (8.08)	85.91 (13.65)	71.39 (11.35)	216.05 (34.34)	191.58 (30.45)	629.17

Note: Figures in brackets in %.
注：括号内数字单位为%。

Mean calcium content and percentage in different species in total calcium amount of plant leaves in subcellular fractions (mg / kg)

不同类型植物叶片钙在亚细胞组分中含量均值及百分比（mg/kg）

Type 分组		Cell Wall 细胞壁	Cytoplasm 胞质	Organelles 细胞器	Total 总量
Calciphile 钙型	Old leaf 老叶	956.36 (66.54)	339.08 (23.60)	141.63 (9.86)	1437.08
	Young leaf 嫩叶	802.92 (59.05)	350.22 (25.76)	206.55 (15.19)	1359.69
Moderate (Karst Region) 中间型（岩溶区）	Old leaf 老叶	562.02 (53.03)	340.26 (32.11)	157.54 (14.86)	1059.82
	Young leaf 嫩叶	447.79 (41.09)	473.83 (43.48)	168.26 (15.44)	1089.88
Moderate (non-Karst Region) 中间型（非岩溶区）	Old leaf 老叶	292.63 (34.95)	346.37 (41.36)	198.35 (23.69)	837.34
	Young leaf 嫩叶	308.55 (35.95)	366.63 (42.72)	183 (21.32)	858.17
Caleifuge 嫌钙型	Old leaf 老叶	277.65 (38.07)	335.53 (46.01)	116.14 (15.92)	729.32
	Young leaf 嫩叶	261.73 (41.60)	284.49 (45.22)	82.95 (13.18)	629.17

Note: Figures in brackets in %.

注：括号内数字单位为‰。

Microstructure of the leaf of *Loropetalum chinense* subjected to calcium-rich karst environment

檵木（*Loropetalum chinense*）叶片的微结构受到富钙岩溶环境的影响

5-8 trees of *Loropetalum chinense* which grow well and have the roughly same age in karst areas and non-karst areas in Maocun, Guilin, were selected to pick up 60 leaves for each tree. The leaves should be mature, healthy and facing sunlight. These leaves were then processed in the laboratory, observed under microscope, counted and photographed. The results showed that: ① the leaf area of *Loropetalum chinense* is reduced in karst areas, while it is thick, compared with that in non-karst areas; ② the upper epidermis, *Loropetalum chinense* in karst areas become thicker, and the upper epidermis cells in irregularly quadrilateral, arranged in crisscross pattern compared with those in non-karst areas; ③ the leaf lower epidermis of *Loropetalum chinense* in karst areas have less stomas, smaller size and higher stoma index, compared with those in non-karst areas.

在桂林毛村岩溶区和非岩溶区，选择生长状况稳定，树龄相当的檵木植株各5-8棵，摘取向阳面中部成熟、健康的叶片各60片，实验室内制片，置于莱卡荧光显微镜下观察，统计并拍照，结果显示：

（1）叶面积、叶厚度：与非岩溶区相比，岩溶区檵木叶片叶面积减小，厚度增加；

（2）上表皮：与非岩溶区相比，岩溶区檵木叶片上表皮厚度增加，上表皮细胞呈不规则四边形，犬牙交错彼此镶嵌；

（3）下表皮：与非岩溶区相比，岩溶区檵木叶片气孔数目较少，尺寸较小，具有较高的气孔指数。

Comparisions of the leaf structure and moisture content of *Loropetalum chinense* between in karst and in non-karst areas
两种土壤上灌木叶片的基础物理参数比较

Area 地区	Leaf length 叶片长度 /cm	Leaf width 叶片宽度 /cm	Leaf thickness 叶片厚度 /cm	Moisture content 总量
Karst Region 岩溶区	3.25	1.49	193.40	51.38
non-Karst Region 非岩溶区	3.34	1.79**	166.20**	53.86

Note: "**" indicates a very significant difference, $P<0.01$, "*" indicates significant difference, no "*" indicates no difference.
注："**"表示差异极显著，$P<0.01$，"*"表示差异显著，无"*"表示无差异。

Loropetalum chinense grown in the field
生长在野外的檵木

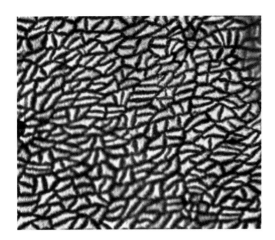

Upper epidermis cell of *Loropetalum chinense* in non-karst areas (50×)

非岩溶区檵木叶片上表皮细胞（50×）

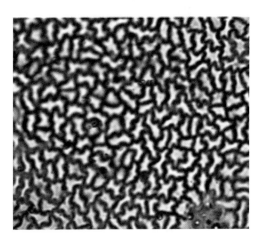

Upper epidermis cell of *Loropetalum chinense* in karst areas (50×)

岩溶区檵木叶片上表皮细胞（50×）

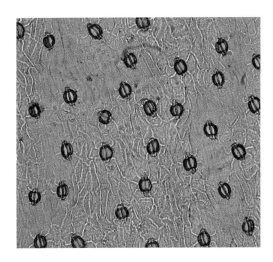

Lower epidermis stomata of *Loropetalum chinense* in non-karst areas (100×)

非岩溶区檵木叶下表皮气孔（100×）

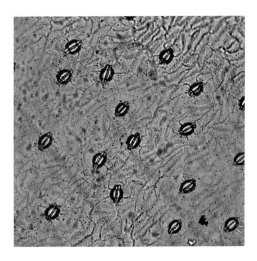

Lower epidermis stomata of *Loropetalum chinense* in karst areas (100×)

岩溶区檵木叶片下表皮气孔（100×）

Preliminary study on adaptation mechanism of honeysuckle in karst environment
初步揭示了忍冬属植物对岩溶环境的适应机制

Honeysuckle grown in the karst Mashan, Guangxi
生长在广西马山石旮旯地上的金银花

Honeysuckle grown in the karst Anlong, Guizou
生长在贵州安龙县石旮旯地上的金银花

Field observations and indoor cultivation experiments were used to study the physiological properties of honeysuckle that grows in Nongla Karst Experimental Site, Guangxi, which provides scientific support to examine the mechanism of suitable plants growing in karst, and the possibility of growing honeysuckle in large scale. Studies have found that :

(1) To adapt to the karst dry environment, the structure of Woodbine leaves demonstrates a series of xerophyte's structural characteristics.

(2) Pot experiments were conducted to assess the response mechanism of honeysuckle to karst drought in terms of photosynthetic physiology and biochemistry after being added 15 mmol/L and 30 mmol/L $CaCl_2$. It was found that in dry process, honeysuckle adapts to inhibit stomatal conductance, slowing down transpiration rate and improving water use efficiency. When the exogenous calcium concentration is 15 mmol/L, photosynthesis of the honeysuckle can be significantly improved, while the too high content of exogenous calcium may cause damage to some honeysuckle.

(3) With laser scanning confocal microscopy imaging, the distribution of calcium ions maintaining the honeysuckle activity in its cells, as well as the way of secreting

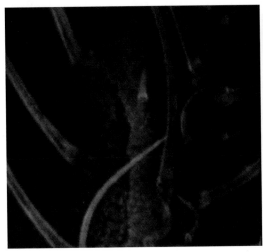 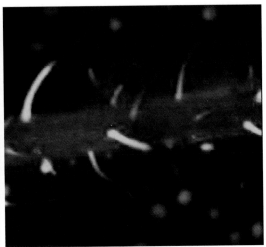

No glands secreting calcium found in the young leaves
叶片幼期尚未出现泌钙腺体

Glands secreting calcium occur when the leaves grow mature
泌钙腺体出现在叶片成熟期

calcium with the gland when it contains excessive calcium (see the photo), while the presence of calcium ions glands can improve its drought resistance by regulating osmotic balance.

(4) By cluster analysis and genetic distance, it was found that the selected materials have a rich genetic diversity and can represent environmental characteristics of different regions, while the calmodulin gene differences affect their ability to adapt to karst environment.

通过野外观测和室内栽培实验相结合的方式,我们研究了生长在广西弄拉岩溶观测站中的忍冬属植物的生理特征,从而为研究岩溶植物适生机制和发展金银花产业提供了理论支持。研究有如下发现:

(1) 忍冬属植物为适应岩溶干旱环境,其叶在形态结构上表现出一系列旱生植物结构特点。

(2) 通过盆栽实验研究了忍冬属植物在添加 15 mmol/L 和 30 mmol/L $CaCl_2$ 后在光合生理生化上对岩溶干旱作出的响应机制。在岩溶干旱过程中,忍冬属植物通过利用土壤中的钙来抑制气孔导度、减慢蒸腾速率以及提高水分利用效率和光合速率来适应岩溶干旱环境。在外源钙浓度为 15 mmol/L 时忍冬属植物的光合作用能得到显著提高,而过高的外源钙则对部分忍冬属植物造成伤害。

(3) 通过激光共聚焦显微镜成像技术,我们认识到维持忍冬属植物生命活动的钙离子在其细胞内分布状况以及钙元素过多时的腺体分泌方式(照片)。钙离子腺体的存在通过调节渗透平衡,提高其抗旱性。

(4) 通过聚类分析和遗传距离对比发现研究所选用的材料具有丰富遗传多样性,能够代表不同区域环境特征。而钙调素基因的差异则影响其对岩溶环境的适应能力。

Karst Dynamic System and Sustainable Development
岩溶动力系统与可持续发展

Integrating control to rocky desertification in karst areas, Southwest China
中国西南岩溶区石漠化综合治理的对策

Karst rocky desertification in Pingguo county, Guangxi.
广西平果县石漠化景观。

Rocky desertification is the extreme degradation of karst ecosystem. Karst ecosystem is fragile, irrational human activities may destroy surface vegetation, produce serious soil erosion, and result in exposure of large areas of bedrock.

Rocky desertification has not only destroyed the human's living and production environment, but also led to poverty in the local areas and constrained the local socio-economic development. To address this issue, the Chinese Government has launched the program of comprehensively controlling rocky desertification in karst areas, which will be implemented during the period of 2006–2015.

According to the characteristics of karst ecosystem and rocky desertification distribution areas in Southwest China, the integrating controls can be summarized as: ① water resources are the fountain head; ② soil resources are the key issue; ③ economic plants are the fundamental; ④ win-win both ecology and economy is the ultimate objective. At the same time, local conditions should be taken into account in implementing the program. IRCK also made a demonstration model of rocky desertification control.

岩溶石漠化是岩溶生态系统退化到极端的表现形式，受人为活动干扰，地表植被遭受破坏，造成土壤侵蚀程度严重，基岩大面积裸露。

石漠化的发生不仅破坏了人类生存、生产的环境，更导致区域的人口贫困，制约了经济社会发展。为此中国政府启动了"岩溶区石漠化综合治理工程项目"，该项目执行期为2006-2015年。

根据中国西南岩溶生态系统特征、石漠化分布的特点，其综合治理的对策可归纳为：水是龙头，土是关键，植被（经济植物）是根本，区域生态经济双赢、农民脱贫致富是目标。在具体实施过程中特别强调因地制宜、分类指导。

The fountain head of water-conjunction on exploitation and utilization of surface and groundwater, using karst watershed as a unit
水是龙头——以岩溶流域为单元，地表、地下水综合开发利用

The karst areas in Southwest China are located in tropical, subtropical monsoon climate, and enjoy abundant moisture and heat conditions. However, due to the bistratal layer of karst hydrogeological structure, the water resources endowed in epikarst zone take up only 8% of the total water resources. As 43%, 83% and 66% of renewable water resources in Yunnan, Guizhou and Guangxi, stored in groundwater systems. For example, the capacity of karst groundwater resources in Guizhou is estimated to be $386.26 \times 10^8 \, m^3/a$, accounting for 80.58% of the total groundwater resources in Guizhou. Nonetheless, due to the heterogeneity of karst aquifers, the groundwater is often deeply buried, which makes it difficult to develop and utilize. It needs multiple methods to exploit water resources in karst areas, such as storing, diverting, pumping and blocking.

To achieve effective exploitation of water resources, the following efforts should be taken: ① to improve the vegetation coverage rate, enhancing the capacity of epikarst zone in regulating and storing water, linking the field water tanks and water ponds with the epikarst spring, and make full use of runoff, especially in the karst areas with deeply ground water; ② to understand the geological and geographic conditions, to find good ways to develop and utilize groundwater, and to try to make water coincide with soil resources, as well as to pay attention to preventing contamination of groundwater aquifer.

西南岩溶区地处热带、亚热带季风气候区，水热条件充足，但双层岩溶水文地质结构，使赋存表层岩溶带的水资源仅为总水资源量的8%，滇、黔、桂可再生水资源分别有43%、83%、66%流入地下水系，水资源以地下水资源为主。如贵州岩溶地下水资源量$386.26 \times 10^8 \, m^3/a$，占全省地下水资源量的80.58%。由于岩溶含水介质的不均一性，地下水通常深埋，开发利用条件差。在岩溶区水资源开发利用要有多种形式，通过蓄、引、提、堵等方式，有效开发水资源。

同时需要注意以下几点：①关注植被覆盖率，提高岩溶表层带对水循环的调蓄能力，注意地头水柜、水窖建设要与岩溶表层带（泉）有机结合，充分利用坡面径流，尤其是地下水埋深大的岩溶区；②充分利用有利的地质、地理条件，开发利用地下水，使水、土资源协调。同时关注地下水与土地利用之间的关系，防止地下水源的污染。

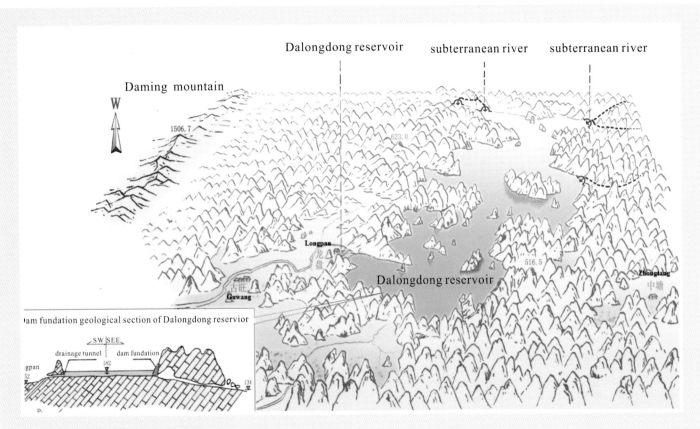

Dalongdong Reservoir in Shanglin County, Guangxi (the reservoir composed surface profile and subterranean stream)

广西上林县大龙洞水库（地下河开发）

（莫日生 提供 From Mo Risheng）

The key issue of soil-saving soil resources, control of water and soil loss, and improvement of land productivity
土是关键——抢救土壤资源、遏制水土流失，提高土地生产力

Demonstration site to soil conservation in Longdong, Guanling County, Guizhou.

贵州省关岭县的岩溶区龙洞小流域水土保持治理关注了岩溶区水土资源的相互配合。

Saving water resources mainly includes three aspects: ① to maintain the total soil volume and protect existing cultivated land. The main measures contains constructing terrain wall to protect soil, making water tanks to water conservancy and improving the function of soil; ② to maintain and improve soil organic matter content, which is known as an important measure in improving soil quality. Under natural conditions, although the limestone soil layer is thin, its organic matter content and soil fertility level is high, organic matter structure remains stable, soil fertility lasts long and soil aggregate structure is good. Therefore, maintaining and improving organic matter content in limestone soil can also increase the content of effective nutrients in limestone soil; ③ to improve the rate of soil formation in limestone area. All the way that can improve the concentration of CO_2 in the soil environment, the concentration and activity of organic acids and soil formation enzymes (especially carbonated glycosidase) can be used to improve the rate of soil and carbonate rocks dissolution.

抢救水土资源主要包括3个方面的内容：①保持土壤总量、保护现有耕地面积。采取的主要措施有坡改梯、砌墙保土、配套小型水利工程，遏制水土流失；② 提高和保持土壤有机质含量是土壤改良的重要措施，在自然状态条件，石灰土土层薄，但有机质含量、土壤肥力水平较高，有机质结构稳定、土壤肥力持久，团粒结构良好。保持和提高石灰土有机质含量对石灰土中若干营养元素有效态含量的提高大有裨益；③提高石灰岩地区成土速率，凡是可以提高土壤环境中的 CO_2 浓度、有机酸和土壤生物酶（尤其是碳酸苷酶）的浓度和活性的技术，均可促进碳酸盐岩的溶蚀和成土速率。

The fundamental of plants (economic plants)-adaptable plant species selection and their nutrients biogeochemical cycles is a fundamental approach to promote the ecological and economic development in karst rocky desertification-affected areas
植物（经济植物）是根本——物种选择、生物地球化学的研究是岩溶石漠化区生态经济建设的途径

Plant species that grow in karst environment have adapted to the karst environment for a long time, and thus a number of unique karst vegetation have been developed. The karst vegetation is characterized by richness in calcium, xerophyte and lithophyte. Previous studies have shown that shrubs and grasses have advantages of growing in karst areas, a large number of shrubs can grow well on rock crevices and fissures. During the growth process, the shrub of Leguminosae needs to absorb a lot of calcium for nutritional purpose. As a result, shrub and other plants play an important role in promoting the ecological and economic development of the karst rocky desertification-affected areas. A few eco-economic species can well adapt to the karst environment in rocky desertification-affected areas, including forage shrubs(e.g., pigeonpea, Tephrosia, woody Zenia) and medicinal vines (e.g., honeysuckle and Quisqualis indica). However, the research and development in this field is relatively weak.

生长在岩溶环境中的植物物种长期对岩溶环境的适应，形成一批岩溶环境中特有的岩溶植被。岩溶植被具有3个基本的特征：富钙性、旱生性、石生性。已有的研究显示灌木、草，尤其是灌木在岩溶区具有生长优势，而大量的灌木则可以在岩石缝隙、石旮旯地上生长良好，且豆科灌木在生长过程吸收大量的钙素作为其营养需要，因此灌木类植物在岩溶石漠化区生态经济建设中将发挥重要的作用。对岩溶石漠化区具有良好适应的生态经济物种，包括饲料灌木木豆、山毛豆、木本任豆；药用藤本植物金银花、使君子等。目前该方面的研究、开发工作相对薄弱。

上图 / Top
Honeysuckle grow on rock crevices in Mashan, Guangxi.
广西马山生长在石旮旯地金银花。

下图 / Bottom
Pigeonpea grow in rocky areas Huanjiang, Guangxi.
广西环江生长在石旮旯地中的木豆。

Ultimate objective achieving win-win situation both ecology and economy
生态经济双赢、农民脱贫致富是目标——因地制宜开展农村产业结构调整

Karst ecosystem in Southwest China, impacted by the type, structure, constituents of carbonate rocks. The water cycle, water resources, soil formation and soil resources utilization rely on the characteristics and operation pattern of karst ecosystem, therefore, it is necessary to divide the karst rocky desertification-affected areas into different types in line with the characteristics of karst ecosystem in southwest China, as well as the regional natural and socio-economic conditions, and rocky desertification situation. As different types of zones have different structural characteristics, different measures should be taken.

The following principles should be taken into account in adjusting the agriculture structure in karst rocky desertification-affected areas in southwest China:

(1) Focusing on key areas. Karst rocky desertification control efforts should focus on the moderate light rocky desertification areas, while the severe rocky desertification areas should be enclosed for afforestation.

(2) Water and soil conservation: to change the traditional industrial structure to pay more attention to the development of economic shrub growth and make livestock.

(3) Acting depending on local conditions: ① It is important to make science-based plan according to the local karst environment: different geological, geomorphological and climatic conditions; ② The specific agriculture operation models should be aligned with local situation.

(4) Sustainability: ① Sustainability means to find an extension of the industrial chain, added-science and technology value and eco-friendly products; ② It is necessary to explore and promote the "companies + production base + farmers" management model and other adaptive management pattern, based on the rural residents diversity and scattered distribution characteristics in karst areas, so that the agriculture productivities can be sustainable in both ecology and economy.

西南岩溶生态系统特征、运行规律受到碳酸盐岩类型、结构、成分和组合特征的影响，如石灰岩、白云岩为物质基础构成的生态系统中的水循环及水资源赋存、土壤形成及土壤资源利用均存在差异。因此，因地制宜开展农村产业结构调整首先要根据西南岩溶发育的特征，结合区域自然、社会经济特点、石漠化状况，对西南岩溶石漠化区进行类型区划分，不同的类型区具有不同结构特征，其农村产业结构调整也应有不同的措施方案。

对西南岩溶石漠化区农村产业结构调整的原则主要有以下考虑：

(1) 重点突出，岩溶石漠化综合治理应明确重点区域，即综合治理的重点是轻度、中度石漠化区，是农村产业结构调整的重点区域，

重度石漠化以封育为主。

(2) 保土节水，改变以粮食生产的传统产业结构，重视发展经济灌木产业和草食畜牧业。

(3) 因地制宜，主要有两层含义：其一，要根据岩溶环境特征，制定合理科学的规划，因地质、地貌和气候条件的差异形成分类指导；其二，产业结构的具体内容和运作模式要与当地特点相吻合。

(4) 可持续性，包括两层含义：其一是产业结构链的延伸、科技附加值的增值和生态友好；其二是根据岩溶环境的分散多样性特点，探索和推广"公司＋基地＋农户"等多种适应性的管理模式，使农村产业结构在生态、经济两方面都具有可持续性。

In Qinglong County, Guizhou Province, the local people have changed their original agriculture structure instead of simply growing corn. Fruit trees (peach) are grown alternatively, under which forage grass is planted to raise sheep. Sheep manure is used as fertilizer during day time, and as biogas at night to solve the energy problem in the rural areas.

贵州省晴隆县改变原来单纯种植玉米的农业结构方式，将玉米改为果树（桃树），桃树下播种牧草，牧草用来养羊，白天羊粪用来肥田，夜晚（圈宿）羊粪用来发酵产沼气解决农村能源。

Characteristics and control of water and soil in karst areas
岩溶区的水土流失特征与防治

Measures to control water and soil loss in karst areas are proposed according to geological, geomorphological and ecological niche conditions.

On the upper slopes, as the soil cover is less than normal, the epikarst zone well developed, and the water cycle is dominated by vertical migration. Water and soil erosion is mainly through vertical infiltration, and engineering measures are largely ineffection. Instead, biological measures should be used in this section: closing hillsides to facilitate afforestation, and planting shrubs that can generate economic and water conservation benefits and have well-developed root systems on rocky areas with soil. These biological measures can increase the surface and underground biomass while promoting the rate of soil formation. Well-developed root systems can effectively keep the soil in-situ, and increase the rate of soil formation with litters.

On the lower slopes, the hillsides have soil cover of different thicknesses are well connected, and are usually cultivated by the local people. When the soil layer completely covers the underlying bedrock, a continuous and complete soil cover is made available. In places which are thickly covered by engineering measures such as soil conservation using walls can be taken to protect soil. In places with thin soil cover, biological measures can be applied, such as growing honeysuckle (*Lonicera japonica*), to maintain soil in-situ.

At the bottom of depressions, soil and water leak into the underground river through sinkholes or shafts. In this case, it is important to prevent soil from running into the subterranean streams. In higher elevation depressions and valley, engineering or biological measures targeting sinkholes can be used to conserve soil.

In low-lying depressions and valleys close to ground water table, the depressions (valleys) are subjected to seasonal flooding. Two main types of measures can be taken: ① Flood diversion canals may be constructed to reduce floods and waterlogging in the depressions. ② Sinkholes may be dredged and their flow dimension expanded if possible so as to improve flood discharge and drainage.

Different water and soil conservation efforts should be applied in different parts of the subterranean stream watershed. In the upper reaches, which are the water supply source, the focus should be on closing hillsides and restoring vegetation. In the runoff area, protective farming may be used, and the above slope and depression (valley) soil conservation measures can be employed. In the discharge areas, which are often large valleys and plains, small and medium-sized reservoirs reservoirs can be built to store flood water during the rainy season. They will help improve water use efficiency and curb soil erosion.

岩溶区水土流失防治的对策要根据地质、地貌及生态位的分区特点，因地制宜地提出防治对策。

坡地上部土壤覆盖偏少，表层岩溶带发育，水循环以垂直运移为主，水土流失也主要以垂直下渗为主，工程措施几乎对此不起作用，如果强行实施工程措施，反而会阻碍正常的水循环途径。在该地段应主要以生物措施为主，一方面封山育林，另一方面在有土的石旮旯地上栽种有经济效益和水保效益、根系发达的灌木。这样，在增加地表、地下生物量的同时，可提高成土速率。发达的根系能有效地保持土壤的原位性，同时能通过枯枝落叶来增加成土速率。

The water table is close to the surface in the Qibainong Depression, resulting the perennial loss of productivity of cultivated lands, Dahua County, Guangxi

广西大化县七百弄洼地与地下水位接近，导致洼地耕地几乎常年丧失生产力

坡地的中下部有不同厚度的土壤覆盖层，连续性好，通常也处于当地居民的耕作范围之中。当土壤层可以完全覆盖下伏基岩时，便具有了较为连续、完整的土壤覆盖。在土壤覆盖层较厚的地方可以考虑砌墙保土等工程措施，防治水土流失；在土壤覆盖层薄的地段则主要考虑土的原位性不受影响，可采取生物措施，如栽培金银花（*Lonicera japonica*）。

洼地底部，水土漏失是通过落水洞或天窗，进入地下河，因此，防止土壤随水流进入地下河成为水土保持的关键。在海拔较高的洼地、谷地，围绕落水洞采用水土保持的工程措施和生物措施，防止土壤随水流进入地下。

海拔较低的洼地、谷地，接近地下水位时，洼地（谷地）易发生季节性涝灾，主要采取2种措施：其一，修建引洪渠，排除洪水及洼地积水；其二，对落水洞进行清淤，有条件的情况下，拓宽落水洞过水断面积，提高排洪、排涝效率。

地下河流域的水土保持主要针对不同的流域部位，采用不同的水土保持对策和措施：上游是补给的源头，重点是封山育林，恢复植被；径流区可采用保护性的农耕，采用坡地、洼地（谷地）的水土对策和措施；排泄区往往是较宽敞的谷地和平原区，可修建中小型水库，用于丰水季节的蓄洪，在提高水资源利用效率的同时，遏制土壤侵蚀。

Dissolution cracks of limestone filled by the surface soil that vertically moving down

垂直下移的地表土壤充填石灰岩溶蚀裂隙

Waterlogging in karst areas
岩溶区的内涝灾害

Karst waterlogging disaster is one of the common eco-environmental issues in karst areas. It often happens in the areas where carbonate rocks are distributed in Guangxi, Guizhou, Yunnan and Hunan in China.

Excessive rainfall is the main reason for the occurrence of waterlogging in karst depressions, and the structural features and input and output conditions of karst depressions are also important factors leading to such disasters.

Damage of forest and vegetation, soil and water loss, rocky desertification, siltation of underground river system and other human activities will exacerbate the degree of karst waterlogging disasters.

To prevent and control karst waterlogging, it is necessary for us to further investigate and better understand the hydrogeological structural characteristics and water cycle process in karst areas.

Karst depressions are the areas where soil resources are concentrated.
They are the major places for local people's living and farming activities. Waterlogging will not only submerge crops, but cause great difficulties to the travelling of local residents

岩溶洼地是土壤资源集中分布的部位，是当地居民生产、生活的主要场所，
内涝发生，不仅淹没农作物，也给当地居民出行带来很大的困难

岩溶内涝灾害是岩溶区普遍存在的岩溶生态环境问题之一。在我国的广西、贵州、云南、湖南等省区的碳酸盐岩分布区都是比较常见的，具有区域性的特点。

过量的降雨是形成岩溶洼地内涝的主要原因，而岩溶洼地系统的结构特征、输入和输出条件也是致灾的重要因素。

森林植被的破坏、水土流失和石漠化、地下河系统的淤塞等人类活动则加剧岩溶内涝的发生，加重岩溶内涝灾害程度。

岩溶内涝的防治需要我们对岩溶区水文地质结构特征和水循环过程调查、了解的程度。

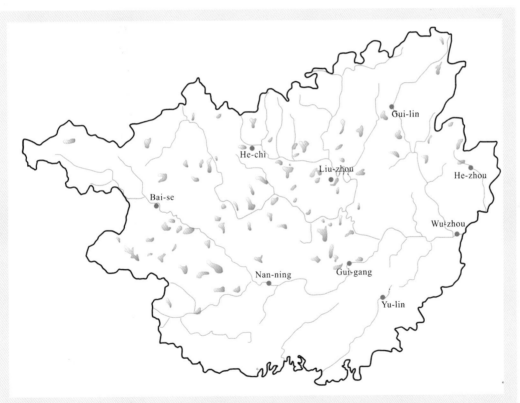

For the counties in Guangxi where more than 30% of their land areas are covered by carbonate rock outcrops, almost each county suffers from karst waterlogging, resulting in an annual average of 61,200 hm^2 of farmland submerged

广西碳酸盐岩出露面积 >30% 的县，几乎每个县都有岩溶内涝灾害，平均每年内涝造成 6.12 万 hm^2 耕地浸没

A karst collapse occurred in residential area in Yulin City, Guangxi
广西壮族自治区玉林市一住宅发生的岩溶塌陷

Collapse risk assessment and early warning in karst area
岩溶区塌陷的风险评价与预警

With the fast growth of urbanization in karst areas, the land, water and mineral resources in these areas have been intensively developed, leading to prominent problems of karst collapse. These problems have become the major geological hazards for cities in karst areas and seriously hindered their construction and development.

Due to the karst collapse features sporadic (temporally), inconspicuous (spatially) and complex (in mechanism) occurrence, it is difficult to be monitored and predicted only with ground-based conventional monitoring tools. The experiment results showed that the change of karst water (vapor) pressure can trigger karst collapse, which can be deemed as the critical conditions for the development of collapse. Therefore, it is possible to predict karst collapse by continuously monitoring the dynamic changes of water (vapor) pressure in karst conduit systems.

A set of research systems and methodologies on the prevention and control of karst collapse

A car got stucked in the mire during a highway collapse in Guangzhou City

广州市一高速公路发生岩溶塌陷，轿车深陷其中

Karst collapse physical modeling experiment

岩溶塌陷物理模型试验

disasters focusing on "formation mechanism model tests, risk assessment and mapping, monitoring and forecasting and information management" have been roughly developed, in which the karst collapse model test, karst collapse geographic information system, karst collapse assessment techniques, and karst collapse monitoring and forecasting techniques have been recognized in China.

随着岩溶区城市化建设的飞速发展，岩溶区土地资源、水资源和矿产资源开发的不断增强，由此引发的岩溶塌陷问题日益突出，已成为岩溶区城市主要地质灾害问题，严重妨碍城市经济建设与发展。

由于岩溶塌陷的产生在时间上具突发性，在空间上具隐蔽性，在机制上具复杂性，因此，普遍认为难以采取地面常规监测手段，对塌陷进行监测预报。试验研究表明，岩溶水（气）压力变化对塌陷具有触发作用，可以以此作为衡量塌陷发育的临界条件。这就意味着通过对岩溶管道系统的水（气）压力的动态变化进行观测，可以达到对塌陷进行预报的目的。

已初步形成一套以"形成机理的模型试验、风险评估与制图、监测预报和信息管理"为特色的岩溶灾害防治研究体系与方法，其中，岩溶塌陷模型试验、岩溶塌陷地理信息系统、岩溶塌陷评估技术、岩溶塌陷监测预报技术等在国内已有一定影响。

Karst geoparks and environmental protection
岩溶区地质公园与环境保护

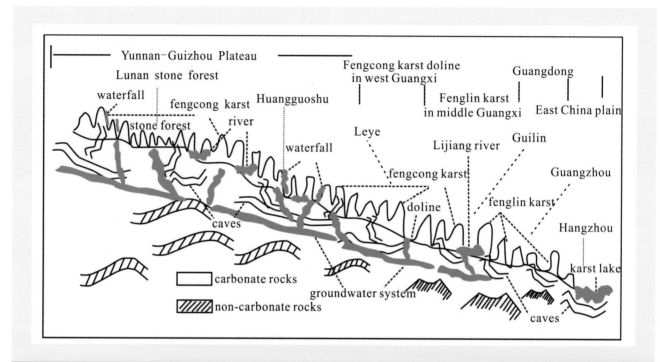

Changes of Karst landscape from East China Plain to Yunnan-Guizhou Plateau
中国华东平原到云贵高原岩溶景观类型变化示意图

In the long geological history, the karst process combines the impacts and changing factors of deposition process, tectonic movement and climate change (including hydrogeological conditions). It directly creates the karst landscape into scenery and is known as a natural sculptor of karst landscape.

So far, 218 national geoparks have been approved by China National Geological Heritage Site (Geo-Park) Review Committee. Among them, 47 geo-parks feature karst landscape as the major part or core landscape.

在漫长的地质历史进程中，岩溶过程将沉积过程、构造运动和气候变化（含水文地质条件）等三者的各种影响和变化因素耦合于一身，是岩溶景观直接的成景过程，是各类岩溶景观的天然雕塑师。

根据国家地质遗迹（地质公园）评审委员会已批准建立国家地质公园218处。其中有47处公园以岩溶景观为主体，或作为核心景观。

	List of part of Geopark related to karst in China 中国与岩溶有关的部分国家公园目录		
Index 序号	Name 名称	Index 序号	Name 名称
1	Shilin (Stone Forest) Geopark 云南石林岩溶峰林国家地质公园	16	Jiuxiang Xiagu Cave National Geopark 云南九乡峡谷洞穴国家地质公园
2	Chongqing Wulong Karst National Geopark 重庆武隆岩溶国家地质公园	17	Shidu National Geopark 北京十渡国家地质公园
3	Dashiwei Tiankeng Group National Geopark 广西百色乐业大石围天坑群国家地质公园	18	Huanglong National Geopark 四川黄龙国家地质公园
4	Xingwen Geopark 四川兴文石海国家地质公园	19	Xiongershan National Geopark 山东枣庄崮山熊耳山国家地质公园
5	Zhijin Cave National Geopark 贵州织金洞国家地质公园	20	Yuntaishan National Geopark 河南焦作云台山国家地质公园
6	Ziyuan Geopark 广西资源国家地质公园	21	Fenghuang National Geopark 湖南凤凰国家地质公园
7	Jiuzhaigou National Geopark 四川九寨沟国家地质公园	22	Anxian Bioherm Karst National Geopark 四川安县生物礁－岩溶国家地质公园
8	Xingyi National Geopark 贵州兴义国家地质公园	23	Fuping Natural Bridge National Geopark 河北阜平天生桥国家地质公园
9	Suiyang Shuanghe Cave National Geopark 贵州绥阳双河洞国家地质公园	24	Pingtang National Geopark 贵州平塘国家地质公园
10	Yimengshan National Geopark 山东沂蒙山国家地质公园	25	Luzhai Xiangqiao Karst Ecological National Geopark 广西鹿寨香桥喀斯特生态国家地质公园
11	Benxi National Geopark 辽宁本溪国家地质公园	26	Baiyunshan National Geopark 河北涞源白云山国家地质公园
12	Fengshan Karst Geopark 广西凤山岩溶地质公园	27	Chenzhou Feitian National Geopark 湖南郴州飞天国家地质公园
13	Wansheng National Geopark 重庆万盛国家地质公园	28	Bagongshan National Geopark 安徽八公山国家地质公园
14	Dahua Qibainong National Geopark 广西大化七百弄国家地质公园	29	Changshan National Geopark 浙江常山国家地质公园
15	Sinan Wujiang Karst National Geopark 贵州思南乌江喀斯特国家地质公园		

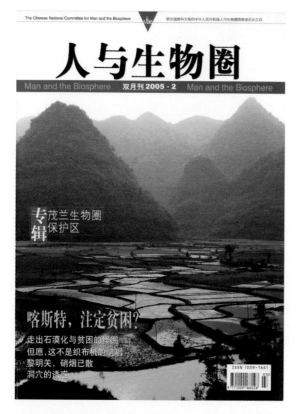

Special edition for Maolan on "Man and the Biosphere"
茂兰生物圈保护区在《人与生物圈》的专辑

Guizhou Maolan Karst Forest Geopark is the largest karst forest ecosystem in subtropical zone, where has a strong original nature and remains stable subtropical vegetation covered in the same latitude. It is an ideal place to conduct qualitative, quantitative and site-based research on karst forest ecosystem. It is also listed by UNESCO into "International Man and Biosphere Reserve Network".

贵州茂兰喀斯特森林地质公园，是地球同纬度地区残存下来的一片面积最大、相对集中、原生性强、相对稳定的喀斯特森林生态系统，是研究喀斯特森林生态特性的天然实验室和难得的定性、定量和定位的研究基地，被联合国教科文组织列为"国际人与生物圈保护区网络"。

(陈向军 提供 / From Chen Xiangjun)

Guizhou Maolan Karst Forest Geopark

贵州茂兰喀斯特森林地质公园——小七孔自然景观

Field Demonstrations on Karst Dynamic System
岩溶动力系统野外研究基地

Guilin Yaji Experimental Site is the first hydrogeological research site in epikarst zones in China
桂林丫吉试验场是我国第一个岩溶表层带水文地质研究的试验场

Yaji Karst Experimental Site was built in 1986 as a Sino-French cooperative project, representing the Fengcong karst spring system in bare karst areas in southern China. It is located 8 km southeast of Guilin City at the border area between Fengcong depression and Fenglin plain area, with a total recharge area of 2 km². It forms a karst hydrogeological system. Its recharge area lies in Fengcong lying areas with 13 low-lying depressions, while its discharge zones is composed of a perennial water spring (No.31 spring) and 3 seasonal spring (No.29, 291 and 32 springs) located in the eastern fringe of Guilin Fenglin Plain. The plain surface has an elevation of 150 m, while the figure for the peak of Fengcong hills area in the recharge area is 650 m, the bottom of its depressions between 250–400 m, and the plain ground only 150 m in eastern and western sides.

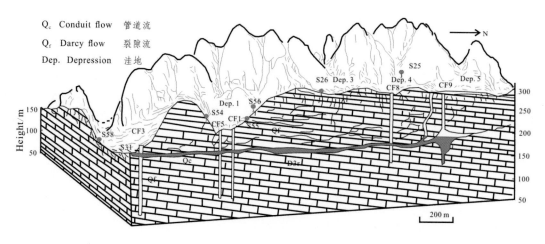

Sketch of geological section of Guilin Yaji Experimental Site
桂林丫吉试验场地地质剖面示意图

桂林丫吉村岩溶试验场作为中法合作项目建于1986年，是中国南方裸露岩溶区具有代表性的峰丛山区岩溶泉域系统。位于桂林市东南郊区8 km的丫吉村附近，处在峰丛洼地和峰林平原的交界地带，总面积2 km²。试验场自成一个岩溶水文地质系统，它的补给区位于峰丛洼地区，有13个洼地，而它的排泄区，由位于桂林峰林平原东部边缘的一个常年流水泉（31号泉）和3个季节性泉（29、291、32号泉）组成。平原面标高150 m，而补给区内的最高峰峰丛山区最高峰达650 m，其内洼地底部标高为250–400 m，而其东西两侧的平原地面标高仅150 m。

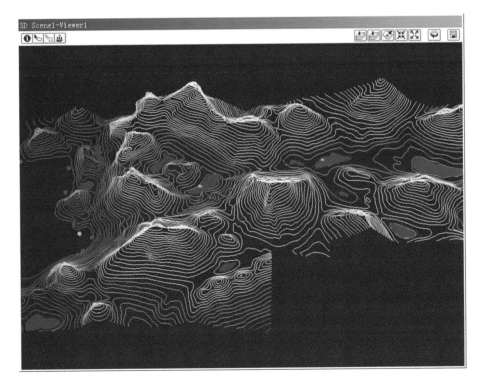

上图 / Top

Experts from France visited the Yaji Experimental Site and guided the build of this site.

法国专家考察丫吉岩溶试验场地，并帮助和指导试验场地的建设。

下图 / Bottom

3-Dimentional view of Yaji Experimental Site.

丫吉试验场地的3维视图。

左图 / Left
Collect data.
采集数据。

右图 / Right
Spring No. 31 and borehole CF1.
S31 号泉及 CF1 钻孔。

Some results from the experimental site:

(1) The terrestrial hydrology linear reservoir model has been introduced to karst groundwater hydrology and has been further developed, shifting the Fengcong karst groundwater model from "black box" to "gray box". Yaji model, a multi-channel peak depression rainwater recharge gap tubes combined vadose zone model, has been developed;

(2) A new set of methods on karst hydrogeological studies has been established, including: application of hydrogeochemical in aqueous medium of karst springs and structure, application of tracer tests in defining spring area recharge boundary, application of computer in fracture analysis, and digitization of aerial photographs and remote sense imaging;

(3) Three storage zones (layers) have been divided in karst Fengcong areas: epikarst zone (upper vadose zone), the lower part of the vadose zone of recharge area, and lower saturation zone in discharge zones;

(4) It provides a site to reveal the operation pattern of karst dynamic system, matter cycle and strengthen the understanding of karst dynamic system;

(5) With the updated field monitoring instruments and equipments, especially by working with Western Kentucky University in US, Nanjing Agricultural University and Huazhong University of Science and Technology in China, we have conducted intensive monitoring on Karst dynamic process and gained new ideas, such as the response process and mechanism of karst dynamic system to storm; carbon transfer and cycle in atmosphere, soil, vegetation and water; and variation of soil dissolved organic carbon, and soil CA enzyme.

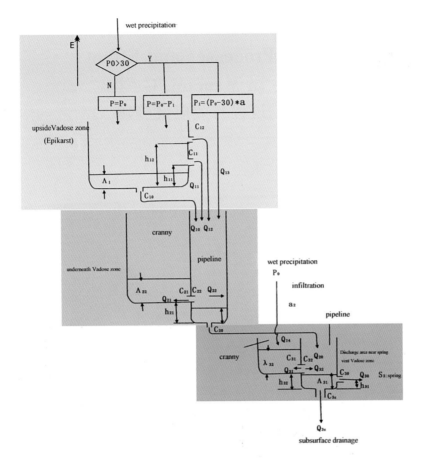

左图 / Left

Tank model of Yaji Experimental Site.

丫吉试验场地的水箱模型。

在桂林丫吉岩溶试验场取得的主要成果：

(1) 将陆地水文学中的线性蓄水模型方法引入岩溶地下水水文学中来，并加以发展，使以往峰丛山区岩溶地下水模型从"黑箱"成为"灰箱"。首次建立了峰洼雨水多途径补给隙管组合包气带模型——丫吉模型；

(2) 建立了一套新的岩溶水文地质方法研究，其中包括：应用水文地球化学理论研究岩溶泉及含水介质结构方法，应用示踪试验圈定泉域补给边界的方法，应用计算机进行裂隙分析、航片解译结果的数字化方法等；

(3) 划分了峰丛山区岩溶水系统三个调蓄带（层），即表层岩溶带（上部包气带）、补给区下部包气带和径流排泄区下部包气带；

(4) 为岩溶动力系统运行规律及其物质循环特点的揭示提供了野外实验基地，并不断深化对岩溶动力系统的认识；

(5) 随着野外观测仪器的更新、设备的改造，尤其是与美国西肯塔基大学、南京农业大学、华中科技大学的合作，对岩溶表层带的动力过程进行了加密的观测，获得了许多新的认识：如岩溶动力系统对暴雨响应的过程及机理；岩溶洼地大气、土壤、植被、水体中碳迁移、循环规律；土壤溶解有机碳、土壤CA酶的动态特征等。

Karst Carbon Sink Processes and Effect Experimental Site in Maocun, Guilin
桂林毛村岩溶碳汇过程与效应试验场

Maocun Experimental Site
毛村试验场

Guilin Maocun Karst Experimental & Observation Station is located in Chaotian, Lingchuan County, Guilin, about 35 km southeast of Guilin.

There are four reasons to select Maocun Village as the experimental site to study karst carbon sink process and effect:

(1) It has a complete karst underground watershed with a recharge area of about 10 km^2;

(2) Pure limestone and sandstone shale are distributed in this area, it is very easy to

make comparative study on karst environment and non-karst environment, and to effectively control the inputs of exogenous water so as to calculate the contribution of exogenous water on karst carbon sink;

(3) Good vegetation communities are well preserved in the area, and there are a set of complete land use patterns: woodland, scrub, grassland and arable land;

(4) Two different types of soil and two different water sources can be selected to conduct comparative study on karst and non-karst areas and to reveal the mechanism of karst carbon sink process.

桂林毛村岩溶生态试验观测站位于灵川县潮田乡，桂林市东南约35 km，交通便利。选择毛村作为岩溶碳汇过程与效应试验场有四个原因：

（1）具有完整的岩溶地下河流域，流域面积约 10 km²；

（2）具有纯的石灰岩和砂页岩分布，非常便于岩溶环境与非岩溶环境的对比；同时能有效控制外源水的输入量，便于计算外源水对岩溶碳汇的促进；

（3）保存有较好的植被群落，林地、灌丛、草地、耕地等土地利用方式较齐全；

（4）可选择两种不同的土壤类型、两种不同的水源，开展岩溶区－非岩溶区对比性的研究，便于从机理上阐述岩溶碳汇过程。

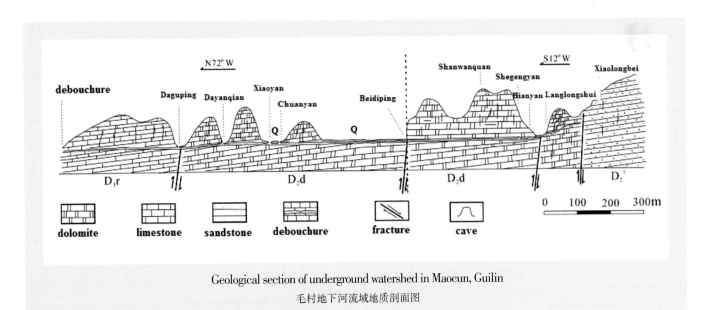

Geological section of underground watershed in Maocun, Guilin

毛村地下河流域地质剖面图

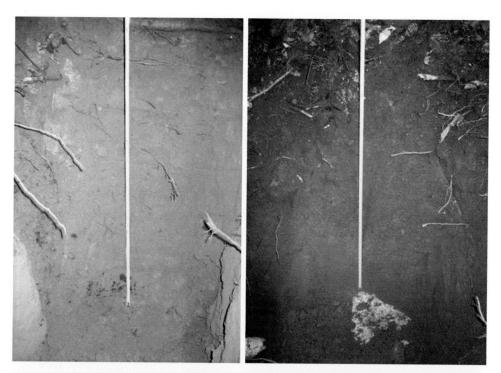

Limestone and sandstone shale red soil profiles in Maocun karst experimental site.
毛村岩溶区石灰土剖面、砂页岩红壤剖面。

Comparison of chemical properties for limestone soil profiles and sandstone shale red soil profiles in Maocun 毛村岩溶区石灰土剖面和砂页岩区红壤剖面的土壤化学性质对比						
Site 分组	Depth/cm 深度/cm	pH	Soil organic carbon/% 土壤有机碳/%	Total nitrogen/% 总氮/%	Total phosphorous/% 总磷/%	Total potassium/% 总钾/%
Karst 岩溶区	10	7.22	3.60	0.63	0.11	0.46
	20	7.34	2.77	0.39	0.08	0.48
	30	7.38	2.87	0.26	0.07	0.49
	50	7.36	2.82	0.18	0.17	0.51
	70	7.45	2.51	0.13	0.10	0.51
	90	4.46	2.08	0.13	0.05	0.51
Non-Karst 非岩溶区	10	4.58	1.56	0.22	0.13	1.05
	20	4.63	1.27	0.13	0.14	0.98
	30	4.69	1.12	0.12	0.14	1.25
	50	4.66	1.27	0.09	0.20	1.06
	70	4.67	1.36	0.09	0.24	0.99
	90	4.78	1.45	0.08	0.16	0.83

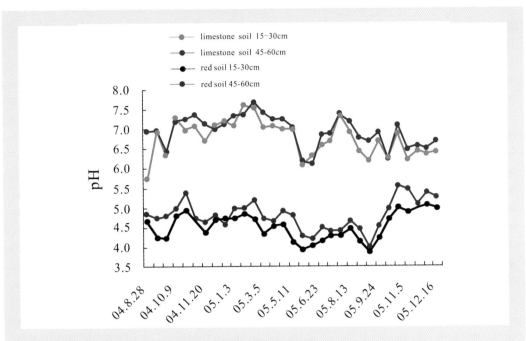

The pH of limestone soil is 2–3 unit higher than of that of red soil

石灰土比红土土壤剖面中土壤 pH 要高出 2–3 个单位

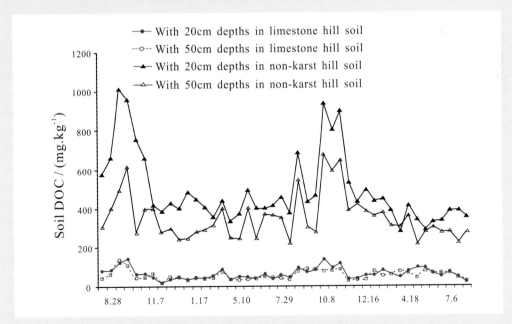

Content of dissolved organic carbon (DOC) in limestone soil is 1/6 of that in sandstone shale red soil

石灰土的溶解有机碳 (DOC) 含量是红壤的 1/6

Experimental Site on Stalagmite Recording Paleoclimate and Modern Cave Drip Monitoring in Panlong Cave, Guilin
盘龙洞洞穴石笋古气候变化记录与现代洞穴滴水监测

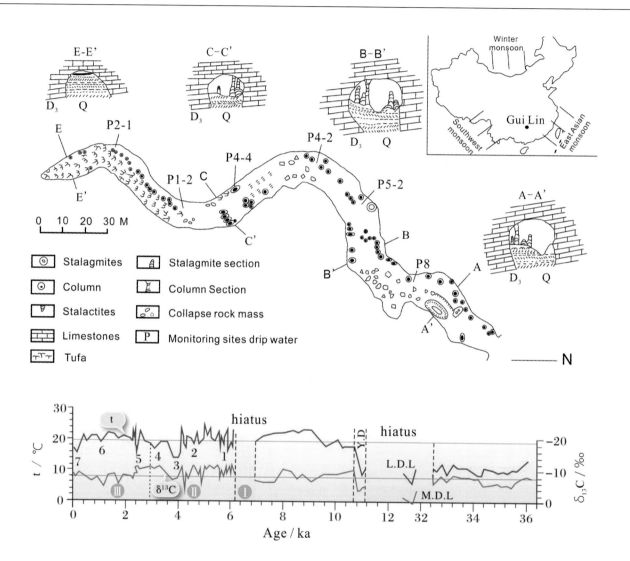

In 1992, the stalagmites in Panlong Cave, Guilin was firstly used to conduct high-resolution paleoclimate change record and reconstruct the climate change pattern over the past 36,000 years.

1992年，利用桂林盘龙洞洞穴石笋率先开展了我国高分辨率古气候变化记录重建研究，构建了3.6万年以来桂林地区气候变化模式。

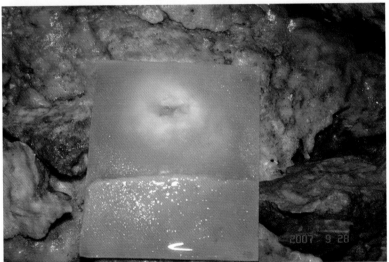

左二图
Left

Receiving device (glass pane) for drip water and speleothem in Panlong Cave.
桂林盘龙洞滴水与化学沉积物收集装置（玻璃板）。

上图
Top

Collection of Stalagmite in Guilin Panlong Cave.
盘龙洞石笋采集。

对页图
Opposite page

Plan of Guilin Panlong Cave and distribution of drip water, speleothem, and meteorological observation sites.
桂林盘龙洞平面图及滴水、沉积物和气象观测点分布。

Cave environmental changes reflect the surface eco-environment and atmospheric changes. Drip water monitory is a good approach to link the changes of surface environment and cave sediment formation. Therefore, dynamic monitoring can provide theoretical basis for paleoclimate reconstruction and interpretation, as well as proposing measures to protect and restore cave resources.

洞穴环境的变化反映了地表生态环境及大气环境的变化，是研究地表环境变化与洞穴古气候研究的桥梁。因此，开展洞穴环境的动态监测，一方面为古气候环境重建解译提供理论依据；另一方面则为洞穴景观资源的保护和修复提出相应的措施。

Experimental Site on Karst Rocky Desertification Control in Karst Fengcong, Guohua, Pingguo County, Guangxi
广西平果果化峰丛洼地石漠化综合治理试验场

Karst Rocky Desertification in Guohua, Pingguo County, Guangxi
广西平果县果化地区岩溶石漠化现象

The demonstration is situated about 15 km from Pingguo County, Guangxi. It has a total area of 6 km^2, with a population of over 1300 people, most of them are of Yao ethnic group. Topographically, it isa typical Fengcong depression, and severely affected by rocky desertification. The demonstration zone was constructed in 2001. According to the spatial distribution patterns of karst Fengcong depression, a three-dimensional eco-agriculture management model and technical system has been proposed, and the following measures have been taken:

At the top of the mountain: closing hillsides for forest conservation, building shelter belts to improve water storage capacity of karst spring system;

At the upper hillside: constructing water ponds and water tanks;

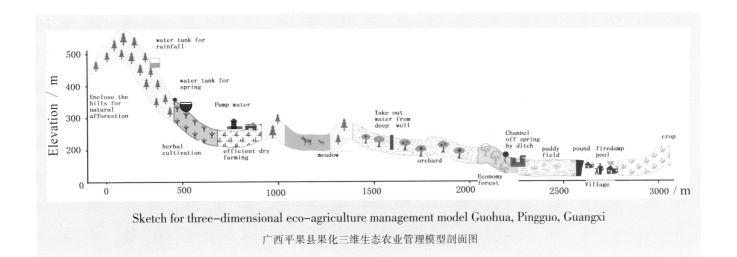

Sketch for three-dimensional eco-agriculture management model Guohua, Pingguo, Guangxi

广西平果县果化三维生态农业管理模型剖面图

At the lower slopes: building economic forest, highly-efficient dry farming land and orchards;
At the foot of hill: growing forage grass to develop animal husbandry sector;
At depression: growing orchards and economic crops.

广西平果县果化示范区距平果县城约 15 km，总面积 6 km²，人口 1300 多人，瑶族。示范区为典型的峰丛洼地地貌，石漠化严重。该示范区的建设始于 2001 年，根据峰丛洼地岩溶地质、水文地质和土壤资源的空间分布格局，提出立体生态农业治理的模式与技术体系，主要采取了以下措施：

山顶：封山育林，构建水源林，提高岩溶表层泉系统的蓄水能力；

山坡上部：修建蓄水池；

山坡中下部：构建经济林、高效旱作农业及果园；

山脚：发展畜牧业；

洼地：发展果园和经济作物。

At the upper hillside: Eight water tanks and water pond have been built, with a total capacity of 3000 m³.

山坡上部：将岩溶表层泉与蓄水水柜的修建结合，共修建 8 个水柜、水窖，总容量 3000 m³。

At the lower part of the hillside: In order to return the broken farmland, Dragon fruit was planted. The experimental results showed that in the crevice, by filling with 5–7.5 kg of soil, two types of dragon fruit trees can be planted and be harvested after two years. It has a survival rate of over 80%. Moreover, the technique to graft high quality dragon fruit has been successfully tested in order to reduce the soil disturbance.

山坡中下部：为了将破碎耕地退下来，探索了石缝地种植火龙果试验。试验结果表明，在裸露的石缝中，每填上 10–15 斤土壤，即可种植火龙果 2 棵，种植 2 年后即可挂果，成活率达到 80%。同时为了减少对土壤的扰动，成功地试验了优质火龙果嫁接技术。

上图 / Top
Water tanks are built at the upper hillside Bottom.
山坡上部修建的水柜。

下图 / Bottom
Dragon fruit is planted at the lower part of the hillside.
山坡中下部种植的火龙果。

At the foot of hills and depressions: the karst area is characterized by shallow soil with sticky texture, which constrains the local agricultural production. In order to improve the limited farmland to increase productivity, soil remediation was conducted: some soil from a nearby non-karst area was taken to increase soil thickness by 10 cm, making the total soil thickness over 30 cm. This can fufill the lowest standard of farming, and improve soil physical structure.

山脚、洼地：岩溶区的土壤具有土壤浅薄、质地黏重等特点，一定程度上制约着当地农业生产，在提高有限耕地生产力的试验中，进行土地整理：从附近土壤层较厚的非岩溶区运土，增加土层厚度10 cm，使总厚度达到或超过30 cm，满足耕作的需求，并增加一定比例的沙土，改善土壤物理结构。

上图 / Top
Increasing soil thickness at the foot of hills and depressions.
山脚和洼地增加土壤厚度。

下图 / Bottom
Harvest of Dragon fruit.
丰收的火龙果。

Photo caption:
Name: Huanglong Geology Park
Location: Sichuan
Inscribed as a UNESCO Natural World Heritage Site: 1992
Listed as a Geopark: 2004
Summary: Huanglong is renowned for its beautiful mountainous scenery, with relatively undisturbed and highly diverse forest ecosystems, combined with the more spectacular localized karst formations, such as travertine pools, waterfalls and limestone shoals. Its travertine terraces and lakes are certainly unique in all of Asia.

照片说明:
名称: 黄龙争艳池
所在地点: 四川
列入联合国世界遗产地时间: 1992 年
列入地质公园时间: 2004 年
概述: 黄龙沟风景名胜区位于四川省阿坝藏族羌族自治州松潘县境内, 面积 700 平方公里。公园以五彩斑斓的钙华池、光芒万丈的雪山、千古沉积的幽幽峡谷、神秘幽静的大森林著称于世, 其中尤以高山彩湖、叠瀑为主的石灰华岩溶景观令人叹为观止。

Chapter 4
Academic Meetings

第四章　学术会议

Since its establishment, IRCK has actively sponsored, hosted and co-sponsored six international symposia. IRCK hosted three meetings: IRCK's 2008 Academic Symposium; the IGCP/SIDA 598 International Working Group Meeting in 2012; and the International Symposium on Karst Water under Global Change Pressure, April 2013. IRCK co-sponsored three meetings: the International Symposium on Geology, Natural Resources and Hazards in Karst Regions (Geokarst 2009) in Vietnam, November 2009; IGCP/SIDA 598 Kick-off Meeting and International Conference on Karst Hydrology and Ecosystems at Western Kentucky University, USA in 2011; and China-Africa Forum Water Resources Dialogue in June 2013. IRCK took part in 22 related meetings and international academic conferences, including the 37th, 38th and 40th Sessions of the IGCP Scientific Board; the 38th, 39th and 40th International Association Hydrogeologists congresses. the 2012 International Geological Congress, and the 2012 International Geographical Congress.

国际岩溶研究中心自成立以来，积极发起、主办和协办国际研讨会7次。主办会议3次：2008年开幕式前学术会议；2012年IGCP/SIDA 598国际工作组会议；2013年4月桂林"岩溶资源、环境与全球变化——认识、缓解与应对"国际研讨会。协办国际学术会议4次：2009年越南协办"岩溶区地质、自然资源及灾害"国际研讨会；2011年美国西肯塔基大学IGCP/SIDA 598 "岩溶系统中环境变化与可持续性"项目启动会；2012年IGCP/SIDA 598项目国际工作组年会；2013年7月中非水资源论坛第一次会议。主动参加相关的国际会议22次，其中参加IGCP科学委员会年会3次、IAH会议3次、国际地质大会IUGS 1次、国际地理联合会IGU 1次。

Meetings Hosted by IRCK
主办的会议

IRCK's 2008 Academic Symposium
2008年国际岩溶研究中心成立学术研讨会

Experts from China and abroad attend the first academic symposium of IRCK.
Front row, right to left: Nico Goldscheider, Tang Changyuan, Liu Zaihua.

中外专家参加第一次国际岩溶研究中心学术研讨会。
前排从右到左：Nico Goldscheider，唐常源，刘再华。

A symposium on the development of international karst science and management was held in Guilin, China on December 12, 2008 to coincide with the establishment of IRCK. 16 experts from 7 countries attended. Among them, Prof. Chris Groves, Western Kentucky University, USA; Prof. Ralf Benischke, Institute of Water Management, Hydrogeology and Geophysics, Joanneum Research, Austria; and Prof. Yongxin Xu, Faculty of Natural Sciences, University of the Western Cape, South Africa introduced their experiences with international training courses. Prof. Nico Goldscheider, the University of Neuchatel, Switzerland, and Prof. Andrej Tyc, University of Silesia, Poland, gave reports on developments in karst hydrogeology and reclamation of fragile karst ecosystems.

Experts from home and abroad attended the meeting.
参加会议的国内外专家。

2008年12月12日,在迎来国际岩溶研究中心挂牌成立之际,组织了一次国际岩溶科学进展与相关管理研讨会,研讨会邀请了7个国家16位专家做报告,美国西肯塔基大学 Chris Groves、奥地利水资源管理研究所 Ralf Benischke、南非西开普大学 Xu Yongxin 介绍了国际组织在国际培训方面的经验,瑞士纳沙泰尔大学 Nico Goldscheider、波兰西里西亚大学 Andrej Tyc 则带来了独具特色的岩溶生态环境和岩溶水文地质研究的进展报告。

		International Seminar on Karst	Conveners
1	Yuan Daoxian	Karst Dynamic System	Wang Yanxin Liu Zaihua Nico Glodscheider Tang Changyuan
2	Chris Groves	Mammoth Cave's contributions to international karst hydrogeological training	
3	Yongxin Xu	Hydrogeological training of UNESCO	
4	Andrej Tyc	Participation of the society and non-governmental organizations in reclamation of fragile karst ecosystems in Poland	
5	Ralf Benischke	Postgraduate Training Courses on Groundwater Tracing Techniques 1969-2005 an International Austrian Contribution in the Field of Geosciences Education	
6	Cao Jianhua	Study on character of karst environment and its contribution to government decision-making	
7	Yu Longjiang	Response and driving mechanisms of organisms to the environment in karst ecosystems	
8	Li Qiang	The ecology adaptation of *Flos Lonicerae* in the karst environment and its effects to harnessing rock desertification	
9	Wang Yanxin	Hydrochemical and isotopic evidences of surface water leakage into karst water systems of Shanxi province, northern China	Chris Groves Ralf Benischke Cao Jianhua Xu Yongxin
10	Nico Goldscheider	Hydrogeology of a glacierised karst aquifer system in the Swiss Alps and possible impacts of climate change on the water resources	
11	Qi Shihua	Persistent Organic Pollutants (POPs) in Karst Environment	
12	Jiang Zhongcheng	Adjusting function of ecological environment on epikarst water	
13	Liu Zaihua	A new direction in searching for the atmospheric CO_2 sink: Considering the joint action of carbonate dissolution, global water cycle and the photosynthetic uptake of DIC by aquatic organism	
14	Tang Changyuan	Variations of Heavy metals in the surface water affected by acid mine drainage at a coalfield in Guizhou	
15	Lei Mingtang	Monitoring and forecasting of sinkhole (karst collapse)	
16	Jiang Yongjun	Natural and anthropogenic factors affecting on groundwater quality of a karst underground river system (Nandong, China)	

上表 / Top table

Speakers and topics at the International Symposium, December 2008
2008 年 12 月国际研讨会的报告人、报告题目

International Symposium on Karst Water under Global Change Pressure, April 2013
2013年4月"岩溶资源、环境与全球变化——认识、缓解与应对"国际研讨会

The International Symposium on Karst Water under Global Change Pressure was sponsored by China Geological Survey and the IAH Commission on Karst Hydrogeology, and supported by the Ministry of Land and Resources of the China, Chinese Academy of Geological Sciences, and the Guilin City Government.

IKG and IRCK organized the meeting, and were involved in preparations for the symposium, including established secretariat group and meeting group.

"岩溶资源、环境与全球变化——认识、缓解与应对"国际研讨会由中国地质调查局、国际水文地质学家协会岩溶专业委员会主办，中国地质科学院岩溶地质研究所、国际岩溶研究中心承办，得到国土资源部、中国地质科学院、桂林市政府的支持，承办单位为会议的顺利召开成立了秘书组、会议组，做了大量的前期工作。

The symposium was held in Guishanhuaxing Hotel, Guilin from April 11 to 13, 2013. The symposium was attended by 138 scholars from 13 countries including China, United States, Britain, Germany, Austria, Spain, Hungary, Serbia, Kenya, Bangladesh, and Thailand. The symposium consisted of an opening ceremony, 14 keynote presentations, 45 special presentations, a closing ceremony and field work.

2013年4月11-13日，研讨会在桂林市桂山华星酒店举行，有来自中国、美国、英国、德国、奥地利、塞尔维亚、西班牙、匈牙利、肯尼亚、孟加拉国、泰国等13个国家和地区的138名学者参加。大会分开幕式、主旨报告、分专题报告、闭幕式和野外考察。大会主旨报告14个，分专题报告45个。

China Geological Survey Deputy Director-General Li Jinfa hosted the opening ceremony.

中国地质调查局副局长李金发主持了会议开幕式。

上左图
Top left

Prof. Chris Groves spoke on behalf of the invited international guests.
Chris Groves 代表参加会议的国际嘉宾讲话。

上中图
Top middle

Dr. Wang Min spoke at the opening ceremony on behalf of the sponsors.
汪民先生代表主办方在开幕式上讲话。

上右图
Top right

Mr. Xu Feng gave a welcome speech on behalf of the Guilin Government.
徐峰代表桂林市政府致欢迎辞。

左上图
Top left

Dr. Petar Milanovic, Dr. Wang Min, Prof. Yuan Daoxian and Prof. Jiang Yuchi at the unveiling ceremony for the Geological Research Center on Global Climate Change, China Geological Survey.

在开幕式上，Petar Milanovic、汪民、袁道先、姜玉池见证中国地质调查局全球气候变化地质研究中心成立。

左下图
Bottom left

Parallel sessions were held with the special topics: Climate change and its impact on the karst water cycle; Increasing disturbance on karst lands and their process; the carbon cycle in karst systems; and Efficient management of karst water.

学术研究讨论的分会场，分会场包括气候变化及其对岩溶水资源的影响、人类活动对岩溶资源和环境的影响、岩溶水资源的有效管理和岩溶地区的地质作用与碳循环等议题。

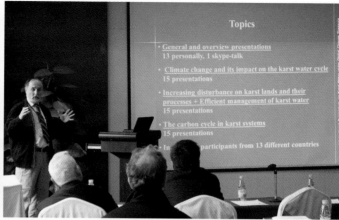

上左图
Top left

Five young researchers won the "Xu Xiake Outstanding Presentation Award". The awardees, pictured with members of the Science Steering Committee, are (left to right): Yang Rui, Huang Fen, Shen Lina, Xiao Qiong, and He Qiufang.

通过研讨会科学指导委员会的推荐、遴选，5位青年参会科技人员获得会议"徐霞客"优秀报告奖，并与科学指导委员会成员合影。
5位获奖青年分别是（从左到右）：杨锐、黄芬、沈利娜、肖琼、贺秋芳。

Prof. Ralf Benischke, representing symposium's Science Steering Committee, made a summary statement. He thought that the symposium showed a deeper understanding of natural karst systems, climate change and the effects of human activity. He was happy to see the support and efforts towards the studies of karst systems and global change that were put forth by China Geological Survey. At the same time, he asserted the relationship between climate change and its effects, and the karst ecological system. Compared with the traditional groundwater hydrogeological model, the Karst Dynamic System model is in a still still under preliminary studies and karst scientists (especially young researchers) need continue making strides in this area.

Ralf Benischke 代表学术研讨会的科学指导委员会做了总结发言，认为此次会议在自然岩溶系统、气候变化和人类活动影响等方面，有了进一步的深入认识，我们也很高兴看到会议的主办方之一——中国地质调查局一直以来对岩溶动力系统与全球变化研究方面给予的很大努力和支持。

同时也提出在气候变化及其影响与岩溶生态系统之间的相互关系；与传统的地下水水文地质模型相比，岩溶动力系统模型（包括水文、地球化学和岩溶生态系统）的构建还处于初始阶段。这就需要岩溶工作者，尤其年轻的岩溶工作者继续前行。

On April 13, experts visited Yaji Karst Experimental Site for the fengcong hydrologic system, and Maocun Karst Experimental Site which is a pilot observation site on karst carbon sink processes in Guilin. They observed the instrumentation used for dynamic monitoring of the hydrologic conditions, and the methodologies were explained.

4月13日，会议安排了桂林丫吉水文地质试验场（主要考察峰丛洼地岩溶水文）、桂林毛村岩溶试验场（主要考察岩溶碳汇过程与对比），展示了岩溶水文、岩溶生态野外研究动态监测的仪器设备、试验布置、技术方法。

During the meeting, the IKG/IRCK exchanged views with the Instituto do Carste, Brazil and Johannes Gutenburg University of Mainz (JGU), Germany.

Photo (left to right, starting fourth from the left): Prof. Augusto Auler, Director of Insitutuo do Carste; Prof. Werner E.G. Müller from Johannes Gutenburg University of Mainz. IRCK/IKG Director Jiang Yuchi; IKG Deputy Director Jiang Zhongcheng.

会议期间，中国地质科学院岩溶地质研究所/国际岩溶研究中心还与巴西岩溶研究所和德国梅恩兹大学就合作项目和合作内容进行交流。

左4：巴西岩溶所所长Augusto Auler研究员，左5：德国梅恩兹大学Werner E.G. Müller教授；右4：岩溶地质研究所副所长蒋忠诚研究员，右5：国际岩溶研究中心主任姜玉池。

左图 / Left

IRCK Executive Deputy Director Cao Jianhua explained the hydrogeological characteristics of Yaji Karst Experimental Site to the visiting experts during the field work.

国际岩溶研究中心常务副主任曹建华研究员给参加野外考察的专家讲解桂林丫吉岩溶试验场的水文地质特征。

IGCP/SIDA 598 International Work Group (IWG) Meeting in 2012
2012年 IGCP/SIDA 598 国际工作组会议

On December 11, 2012, during the IRCK training course, 15 representatives of the IGCP/SIDA 598 International Working Group came to Guilin from China, Slovenia, Spain and the United States for the 598 IWG meeting. The meeting was convened by project leader, Prof. Zhang Cheng, and IRCK Academic Committee director Yuan Daoxian gave a speech. At the meeting, they talked about the development of the project, and discussed the problems of karst area resources and environment under global warming; the challenges of economic and social sustainable development in karst areas; the effects of mining and highway construction in the karst environment; integrating control of rocky desertification in karst areas; protection of karst water resources; fighting drought, flood and collapse in karst areas; and the application

of modern technology in karst dynamic system. Special reports were given by eight experts: Zhang Cheng, Chris Groves, Tadej Slabe, Martin Knez, Bartolomé Andreo-Navarro, and Alena Petrvalská.

2012年12月11日，利用国际岩溶研究中心培训班的机会，邀请了中国、美国、斯洛文尼亚、西班牙等IGCP/SIDA 598国际工作组的代表15人，在桂林召开了598项目年度的国际工作组会议，会议由598项目国际工作组主席章程博士主持，国际岩溶研究中心学术委员会主席袁道先院士做主旨发言。

会议汇报了项目执行以来取得的进展，研讨了面对全球气候变化下岩溶区资源、环境问题，及岩溶区经济社会可持续发展面临的挑战：矿山开采、高速公路的建设等对岩溶环境的影响；岩溶区石漠化治理的途径；岩溶水资源的保护及旱涝灾害、塌陷的发生；现代技术手段在岩溶动力系统研究中的应用。Zhang Cheng、Chris Groves、Tadej Slabe、Martin Knez、Bartolomé Andreo-Navarro、Alena Petrvalská等8位专家做了专题报告。

上左图
Top left

Prof. Bartolomé Andreo-Navarro, karst expert from Spain, gave a report on "Protecting groundwater in karst media".

西班牙岩溶专家Bartolomé Andreo-Navarro做"Protecting groundwater in karst media"报告。

上右图
Top right

Prof. Martin Knez, from Karst Research Institute ZRC SAZU, Slovenia, gave a report on "Planning traffic roads crossing karst".

斯洛文尼亚专家Martin Knez做"Planning traffic roads crossing karst"专题报告。

Meetings Co-Sponsored by IRCK
协办的会议

IRCK chaired two karst symposia at the 34th International Geological Congress, August 2012.
2012年8月，作为岩溶专题召集人，主持第34届地质大会两个会场

The 34th International Geological Congress (IGC) was held in Brisbane, Australia from August 5 to 10, 2012. Nearly 6000 geologists from 137 countries and regions participated in the event which is held once every four years. The congress theme was "Unearthing our past and future– Resourcing tomorrow", and it included presentations (3232 presentations, 37 themes, and 189 symposia), earth science exhibits (283 displays), and field excursions (more than 50 trips).

Prof. Jiang Zhongcheng, Prof. Cao Jianhua, Prof. Zhang Cheng, Prof. Lei Mingtang, Ms. Guo Fang and Dr. Pu Junbing represented the IKG/IRCK at the congress. Moreover, Prof. Jiang, Prof. Cao, and Prof. Zhang convened and coordinated two symposia: T29.2 "Karst: processes, environments and paleoenvironmental records (IGCP/SIDA 598)" and T36.2: "Environmental change and sustainability in karst systems: relations to climate change and anthropogenic activities (2011–2016) [IGCP/SIDA Project 598]". More than 50 scientists from Australia, China, Germany, Indonesia, Romania, Russia, Spain, and the United States participated in the symposia, and 18 of them presented their academic reports.

2012年8月5-12日，第34届国际地质大会（IGC）在澳大利亚布里斯班市会展中心举行，来自全球137个国家和地区的5924名地质科学家齐聚一堂，参加四年一届的"地学奥林匹克"盛会。本次大会以"探讨过去，揭示未来——为人类的明天提供资源"为主题，由室内会议（3232个报告、37个主题、189个分专题讨论会）、地学展览（283个展区）和野外地质考察（超过50条考察路线）三部分组成。

蒋忠诚研究员、曹建华研究员、章程研究员、雷明堂研究员、郭芳副研究员和蒲俊兵博士作为国际岩溶研究中心/岩溶地质研究所代表团成员参加了此次盛会。同时，作为召集人，主持召开了两场专题研讨会："岩溶过程、环境及古环境记录"（T29.2 Karst: processes, environment and paleoenvironmental records）和"环境变化与岩溶系统可持续性，IGCP/SIDA 598项目"（T36.4 Environmental change and sustainability in karst systems: relations to climate change and anthropogenic activities (2011–2016) [IGCP Project 598]），来自美国、德国、中国、罗马尼亚、澳大利亚、俄罗斯、西班牙、印度尼西亚等国的50多名科学家参加，其中18人做了会议学术报告。

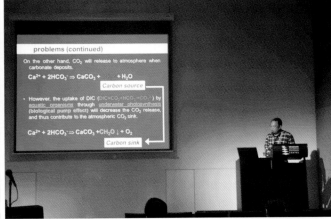

左上图
Top left
Prof. Lei Mingtang gave a report in the T36.4 symposium.
雷明堂研究员在T36.4专题会场做主旨报告。

右上图
Top right
Prof. Liu Zaihua was the keynote speaker for the T29.2 symposium.
刘再华研究员在T29.2专题会场做主旨报告。

左下图
Bottom left
The T29.2 symposium.
T29.2专题会场。

右下图
Bottom right
Plenary of the 34th International Geological Congress.
大会会场。

Geokarst 2009: International Symposium on Geology, Natural Resources and Hazards in Karst Regions, Vietnam

2009年1月，在越南协办"岩溶区的地质、自然资源与灾害"研讨会

The International Symposium on Geology, Natural Resources and Hazards in Karst Regions (Geokarst 2009) was held in Hanoi, Vietnam from November 12 to 15, 2009. The symposium was sponsored by UNESCO, IRCK, Hanoi University of Mining and Geology (Vietnam), University of Natural Resources and Applied Life Sciences (Austria), Sapienza——University of Rome (Italy), National Program KC.08-6-10 (Vietnam), Center for Water Resources Planning and Investigation (Vietnam), Vietnam Institute of Geosciences and Mineral Reources, and International Atomic Energy Agency (Austria). It was organized by Hano University of Mining and Geology, Vietnam.

"岩溶区的地质、自然资源和灾害"国际研讨会，由联合国教科文组织国际岩溶研究中心、河内矿产和地质大学以及来自越南、奥地利、意大利的大学和科研院所等9家单位联合主办，河内矿产和地质大学承办，会议于2009年11月12－15日在越南河内举行。

左上图
Top left

IRCK representatives Prof. Yuan Daoxian, Prof. He Shiyi, Prof. Xie Yunqiu and Ph.D student Kang Zhiqiang attended Geokarst 2009. Prof. Yuan Daoxian chaired a session on "natural hazards in karst regions and their impacts", which was one of the four sections of this symposium. He also delivered a keynote speech entitled "The groundwater protection issue in karst regions of southwest China".

国际岩溶研究中心袁道先院士、何师意研究员、谢运球研究员及康志强博士生等组成代表团参加会议，袁道先在会上做题为"中国西南岩溶区地下水保护"主题报告。

左下图
Bottom left

After the symposium, participants took part in a two-day field excursion to the unique geological diversity of the world-famous karst terrain in the Ha Long Bay-Cat Ba Island area.

会后，代表们参加了为期两天的野外考察，主要考察岩溶台地的地质多样性、世界自然遗产地下龙湾和吉婆岛海上岩溶地貌。下图为世界自然遗产地越南下龙湾景观。

China–Africa Water Forum – Water Resources Dialogue, June 2013 in Cape Town, South Africa

2013年6月协办南非开普敦"中非水资源论坛"第一次研讨会

The first international conference of the China–Africa Water Forum was organized by the Africa Groundwater Academy, with support from IRCK; the Council for Science and Industrial Research (CISR), South Africa; African Ministers' Council on Water (AMCOW) Africa Groundwater Commission; NEPAD Southern African Network of Water Centres of Excellence (SANWATCE); and the UNESCO Chair in Hydrogeology at UWC. The meeting was held at the University of the Western Cape (UWC) from June 26 to 28, 2013. The theme of the meeting was "Sustainable utilization of water resources in developing countries".

"中非水资源论坛"的第一次学术研讨会,由南非地下水科学院、联合国教科文组织国际岩溶研究中心、南非科学技术研究委员会、非洲部长级水资源委员会、联合国教科文组织水文地质教席主办,南非西开普敦大学承办;会议于 2013 年 6 月 26-28 日在南非开普敦西开普敦大学举行;会议的主题是"发展中国家水资源可持续利用"。

A total of 34 attendees from 10 Chinese organizations and 39 attendees from 12 African countries participated in the meeting. Discussion topics included exploration and development of surface and underground water (including karst water), management and sustainable utilization of water resources, comprehensive water resource management, and the operation of the China Africa Water Forum and development of substantial projects. There were three days of academic exchange activities and 28 academic reports, including six keynote presentations. The first Working Group of the China Africa Water Forum was established during the conference, and it was decided that the next activity should be held in China. The picture above is a group photo of the first Working Group.

参加研讨会的代表有来自中国 10 个单位的 34 人,来自非洲 12 个国家的 39 人。会议围绕地表水、地下水(包括岩溶水)的勘探和开发,水资源的管理和可持续利用,水资源管理的综合研究,中非水资源论坛的运作和实质性项目的开展等方面进行研讨。大会安排了 3 天学术交流活动,28 个学术报告,其中 6 个主旨报告。同时成立了"中非水资源论坛"第一期工作组,讨论决定下一期论坛活动在中国召开。照片为部分第一期工作组成员合影。

Representatives of IRCK, Prof. Jiang Yuchi and Prof. Cao Jianhua gave report on karst dynamic system and its application in China and the exploitation, utilization and protection of the groundwater in the karst areas of Southwest China. They also promoted the goals, purpose and recent activities of IRCK.

国际岩溶研究中心姜玉池、曹建华代表国际岩溶研究中心参加研讨会,并做"岩溶动力学及在中国的实践(Karst Dynamics and its application in China)"和"中国西南岩溶地下水资源开发、利用与保护(Exploitation, utilization and protection of the groundwater in the karst area of Southwest China)"报告,同时宣传了国际岩溶研究中心的目标和宗旨及最近的相关活动。

Meetings Attended
参加的会议

37th, 38th, and 40th Sessions of the IGCP Scientific Board
参加 IGCP 科学委员会 37 届、38 届、40 届年会

At the invitation of the IGCP Executive Secretary Robert Missotten, a five-member delegation of IRCK attended to the 37th Session of the IGCP Scientific Board, held February 18-20, 2009 at UNESCO Headquarters in Paris. At the Open Session, Prof. Cao Jianhua delivered a report on behalf of the IRCK Director detailed IRCK's opening ceremony, activities in 2008 and work plan for 2009.

Photo: Prof. Vivi Vajda, Chairperson of the IGCP Scientific Board, with representatives of IRCK.

受 IGCP 秘书处 Robert Missotten 博士的邀请，作为教科文组织第一个地学领域的二类研究中心，国际岩溶研究中心参加 IGCP

科学委员会 37 届年会 (2009 年 2 月 18 – 20 日)，国际岩溶研究中心代表团共 5 人参加会议。会上，曹建华博士代表 IRCK 主任做 IRCK 2008 年进展、成立及 2009 年工作计划的报告。

照片为会后国际岩溶研究中心部分代表与 IGCP 科学委员会执行主席 Vajda Vivi 教授合影。

上左图
Top left

IRCK Governing Board member and director of the Hoffman Environmental Research Institute, Western Kentucky University, Prof. Chris Groves gave a review of the IGCP karst projects and the long-term cooperation between China and the United States. He also supported and congratulated the establishment of IRCK.

IRCK 理事会成员、美国西肯塔基大学霍夫曼环境研究所所长 Chris Groves 教授回顾 IGCP 岩溶项目经历和中美长期合作的过程，表达了对 IRCK 成立的支持和祝贺。

上右图
Top right

IRCK Academic Committee Director Yuan Daoxian spoke about the opening of IRCK and its international work, especially that in support of karst scientific research in developing countries and training in environmental protection of water resources, which received support from the African delegates.

国际岩溶研究中心袁道先理事做了评述性的发言，强调了 IRCK 开放和国际化运行，尤其强调了支持发展中国家岩溶科学研究和水资源环境保护培训，得到在场非洲国家代表的响应。

An IRCK delegation attended both the close session and open session of the IGCP Scientific Board in Paris on February 17–19, 2010. Prof. Cao Jianhua delivered reports on the activities of IRCK in 2009 during both sessions, and received a positive response from members of the IGCP Scientific Board and representatives of countries participating in the sessions. The sense of the sessions was that in the first year after its establishment, IRCK has made remarkable achievements in scientific research, international exchange and cooperation, training and karst science popularization and dissemination, and also made encouraging progress in its administerial settings, organizational management and operational condition. A good start has been made on laying a solid foundation for the realization of the IRCK's objectives.

2010年2月17-19日，国际岩溶研究中心代表团应邀参加IGCP科学委员会第38届年会，中心常务副主任曹建华博士分别在第38届科学委员会闭门会议和开放会议上做了2009年度中心工作报告，得到了委员们和与会各国代表的热烈响应和好评，认为中心成立一年来在科学研究、国际合作与交流、培训与科普等方面取得了显著成绩，在组织机构、行政管理、运行条件等方面均有了可喜进展与改善，良好的开端和运行条件为中心目标的实现打下了基础。

第四章　学术会议

IRCK Executive Deputy Director Cao Jianhua delivered reports on the activities of IRCK in 2009 during the 38th Close Session of the IGCP Scientific Board.

中心常务副主任曹建华博士在第 38 届科学委员会闭门会议上做国际岩溶研究中心 2009 年年度工作报告。

Prof. Dong Shuwen and Prof. Cao Jianhua communicated with Dr. Giuseppe Arduino from the UNESCO Office in Jakarta.

会间，国际岩溶研究中心理事会委员董树文教授、中心常务副主任曹建华博士与教科文组织雅加达办公室 Giuseppe Arduino 博士交流未来合作事宜。

The IGCP 40th anniversary celebration was held at UNESCO Headquarters on February 22, 2012. UNESCO Director-General Irina Bokova and IUGS President Alberto Riccardi delivered the welcome address. IGCP Chairperson Vivi Vajda gave a review of the main research results of IGCP's 40 years.

2012年2月22日，IGCP成立40周年纪念大会在UNESCO总部隆重举行，UNESCO总干事Bokova女士和IUGS主席Riccardi教授致欢迎词，IGCP科学执行局主席Vajda教授总结回顾了IGCP 40年来所取得的主要成果和进展。

Source: http://www.caqs.ac.cn

The International Geoscience Program (IGCP) is a joint operation of the International Union of Geological Sciences (IUGS) and UNESCO. Since 1972 UNESCO and IUGS have worked together through IGCP to combat the challenges of different strata geological processes between global continent. In 2012, IGCP celebrated its 40th anniversary, and nine-persons China's delegation participated in the celebration activities.

Photo: Chinese delegates with Dr. Gretchen Kalonji, UNESCO Assistant Director-General for Natural Sciences (provided by CAGS).

Left to right: Jin Xiaochi, Nie Fengjun, Lian Changyun, Dong Shuwen, Gretchen Kalonji, Jiang Jianjun, Yuan Xiaohong, Wang Wei, Fei Yue, Zhang Cheng.

联合国教科文组织（UNESCO）国际地球科学计划（IGCP）是 UNESCO 和国际地质科学联合会（IUGS）于 1972 年为应对全球不同大陆之间的地层以及地质过程的对比所面临的挑战发起创建的国际地质计划，是联合国系统中唯一的地球科学计划，2012 年迎来 IGCP 40 周年，中国 IGCP 全委会应邀组织 9 人代表团出席纪念活动。

照片为中国代表团会后同 UNESCO 助理总干事 Gretchen Kalonji 合影（由中国地质科学院提供）。

从左到右：金小赤、聂凤君、连长云、董树文、Gretchen Kalonji、姜建军、袁小虹、王巍、费玥、章程。

上图
Top

During the meeting IRCK displayed six posters which introduced its construction and major activities. Prof. Zhang Cheng presented an academic report on "Climate change, water resources and water environment".
Photo: Prof. Zhang Cheng was introducing IRCK to foreign experts and discussing future cooperation.

按照会议安排，国际岩溶研究中心制作了6块展板，介绍了中心的建设情况和主要活动。图为中心秘书长章程向外国专家介绍展板内容及相关合作事宜。在会议期间，章程博士还代表国际岩溶研究中心做了"气候变化、水资源与水环境"学术报告。

第四章　学术会议

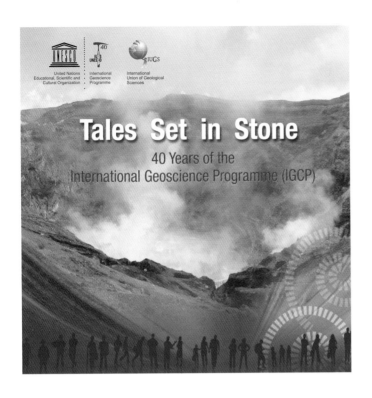

The UNESCO publications celebrating the 40th anniversary of IGCP contained double-page spreads detailed the karst-related IGCP projects.

在UNESCO为IGCP创建40年整理撰写的纪念出版物中，与岩溶相关的IGCP项目获得双倍的页码，用以介绍其取得的新认识、新成果。

Looking Ahead

The future continues to look bright for karst research; much has been learned but questions remain. UNESCO and IUGS partnerships will continue to serve as a leading platform for international communication in karst science, both by way of IGCP 598: Environmental Change and Sustainability in Karst Systems (2011-2015) and the International Research Center on Karst (IRCK). While the countries that have most strongly supported the IGCP karst projects (including China, Slovenia, Spain, and the USA) and continue to do so, interest continues to grow. IGCP 513, which ended in 2010, attracted active participation from 44 countries, and IGCP 598 has co-leaders from Asia, Europe, and North America and, for the first time, the southern hemisphere (Brazil). IGCP 598 has also been awarded supplementary support from the Swedish International Development Agency (SIDA) in recognition of its training courses.

The IRCK also continues to grow as it meets 21st century challenges with excellent facilities at the Institute of Karst Geology in Guilin. Principal financial support comes from the Chinese government so that Chinese administrative leadership comes together with international scientific leadership; present members of the academic committee of the IRCK represent 13 countries. Current plans envisage a rise in the staff of the IRCK to 60 by 2020.

We and our successors expect to be able to report additional successes at IGCP's 50th and perhaps even its 75th anniversary celebrations!

Chris Groves, *Hoffman Environmental Research Institute, Western Kentucky University, USA;* **Yuan Daoxian** *and* **Zhang Cheng**, *International Research Center on Karst under the Auspices of UNESCO, China and Institute of Karst Geology, Chinese Academy of Geological Sciences, China*

IGCP 299, 379, 448, 513, 598: Global Efforts to Understand the Nature of Karst Systems: over two Decades with the IGCP (1990-2015)

国际岩溶研究中心 6'年历程

The 38th IAH Congress, Poland and the 39th IAH Congress, Canada
参加 IHA 会议，波兰 38 届年会，加拿大 39 届年会

IRCK representatives Prof. Cao Jianhua and Dr. Li Qiang attended the 38th IAH Congress in Krakow, Poland on September 10-12, 2010. Approximately 600 representatives from 35 countries participated in the congress, and 521 papers were published in the conference proceedings. The theme of the congress was "Groundwater sustainable development", with six symposia: ① groundwater quality sustainability; ② groundwater and dependent ecosystem; ③ aquifer management; ④ mineral and thermal water; ⑤ data processing in hydrogeology; and ⑥ general hydrogeological problems.

第 38 届 IAH 大会于 2010 年 9 月 10-12 日在波兰克拉科夫举行，来自世界各地共 35 个国家近 600 名代表参会，会前收录论文 521 篇。此次会议以"地下水可持续发展"为主题，设立了地下水水质的可持续性、地下水生态系统、含水层管理、矿山与深部水循环、水文地质数据处理以及水文地质中的常见问题等六个专题活动。

IRCK delegation Prof. Cao Jianhua and Dr. Li Qiang attended the congress, where they learned that eco-hydrology is the primary direction for current developments in hydrogeology. The meeting highlighted new topics relevant to proper groundwater utilization and protection in the karst regions of Southwest China. More hydrogeological date acquisition methods were learned by communicating with Wilhelm Struckmeier, IAH President, which paved the way for IRCK karst hydrogeological mapping programmes.

After the congress, IRCK Academic Committee member Prof. Andrej Tyc from the University of Silesia hosted the IRCK delegates to visit the University of Silesia and investigate the typical karst landscape in Poland.

曹建华研究员和李强博士代表国际岩溶研究中心参加了会议，认识到生态水文学是今后水文地质学的发展方向，从而为中国西南岩溶地下水的合理开发和保护研究提出新的课题；通过与 IAH 主席 Wilhelm Struckmeier 教授的交流，为编制全球岩溶分布图获得基础图件来源。

会后在国际岩溶研究中心学术委员会委员波兰西里西亚大学 Andrej Tyc 教授的陪同下，参观了西里西亚大学、考察了波兰典型岩溶。

The 39th IAH Congress was held in Niagara Falls, Canada on September 14–28, 2012 and was organized by the IAH Canadian National Chapter. More than 900 hydrogeologists from 80 countries and regions attended the conference and discussed problems in groundwater. Prof. Yuan Daoxian, Prof. Xie Yunqiu, Prof. He Shiyi and Mr. Wang Jinliang participated in the conference as IRCK/IKG representatives.

Photo: Field trip to Bruce Peninsula.

第 39 届 IAH 大会于 2012 年 9 月 14 – 28 日在加拿大尼亚加拉瀑布城召开，来自世界 80 个国家和地区的 900 多名水文地质学家齐聚一堂，共同探讨地下水问题。袁道先院士、谢运球研究员、何师意研究员和汪进良助理研究员代表国际岩溶研究中心 / 岩溶地质研究所参加会议。

照片为国际岩溶研究中心代表团参加会间野外考察——尼亚加拉瀑布上游地区 Bruce 半岛地质。

IRCK/IKG representatives had a talk with the groundwater monitoring equipment exhibition booth of a Dutch company during the 39th IAH Congress

会议期间，国际岩溶研究中心代表参观和询问最新水文动态监测仪器设备。

A hydrogeology field trip after the meeting was arranged on September 21-24. Led by Prof. Derek C. Ford, the group visited the karst of Flowerpot Island.

Photo (left to right): Zeng Haitao, Liu Zaihua, Yuan Daoxian, Petar Milanovic, Derek Ford, Wu Aiming, Ralf Benischke, Shen Licheng.

会后，由国际岩溶研究中心理事会委员 Derek Ford 教授带队，考察花瓶岛水文地质。

从左到右：曾海涛、刘再华、袁道先、Petar Milanovic、Derek Ford、吴爱明、Ralf Benischke、沈立成。

IRCK Participated in the 32nd International Geographical Congress
参加德国第 32 届 IGC 年会

The 32nd International Geographical Congress (IGC) was held on August 26–30, 2012 at the University of Cologne, Germany. Nearly 3000 people from more than 60 countries attended the meeting. The congress featured four main topics: Global change and globalization, Society and environment, Risks and conflicts, and Urbanization and demographic change. More than 30 IGU scientific committees convened 400 academic seminars and 180 report posters. Prof. Cao Jianhua, assistant researcher Ms. Huang Fen and Ms. Lu Qian attended the conference as representatives of IRCK/IKG.

Delegation of IRCK/IKG and researchers from other countries took photo together during the congress.

2012年8月26-30日，第32届国际地理大会（IGC）在德国科隆大学举行，来自全球60多个国家和地区的近3000人出席了这次会议，4个主要主题是：全球变化与全球化（Global change and globalisation），社会与环境（Society and environment），风险和冲突（Risks and conflicts），城市化和人口结构变化（Urbanisation and demographic change）。IGU所属30多个委员会组织了专题学术交流。国际岩溶研究中心/岩溶地质研究所曹建华研究员、黄芬助理研究员、卢茜助理研究员参加了此次盛会。

照片为国际岩溶研究中心代表参加IGC会议，代表团与各国学者代表合影。

After the meeting the group visited the karst area in Iserlohn, organized by Prof. Martin Trappe from Catholic University of Eichstaett-Ingolstadt, the local organizer of the Karst Commission of the International Geographical Union (IGU) in Germany.

会后在德国艾希施泰特天主教大学（Catholic University of Eichstaett-Ingolstadt）马丁·特雷普教授（Martin Trappe）的带领下，考察了伊瑟隆岩溶区（Iserlohn karst area），黑摩尔国家石上森林（National Geotope Felsenmeer at Hemer），参观了代兴赫勒洞穴（Dechenhöhle）和伊瑟隆德国洞穴博物馆（Deutsches Höhlenmuseum Iserlohn）。

15th International Congress of Speleology, USA
参加美国第 15 届 ICS 年会

The 15th International Congress of Speleology (ICS), themed "Karst Horizons", was held at Schreiner University in Kerrville, Texas, USA, from July 19 to 26, 2009. IRCK representatives, submitted seven academic papers to the congress and published in proceedings.

Photo: IRCK Secretary–General Prof. Zhang Cheng gave a presentation "Land uses impact on karst processes".

2009 年 7 月 19–26 日,由国际洞穴联合会主办,美国国家洞穴协会承办的第 15 届国际洞穴大会,在美国得克萨斯州克维尔镇施赖纳大学召开。

国际岩溶研究中心代表章程研究员参加了会议,会前国际岩溶研究中心向大会提交了 7 篇学术论文,均收录在会议出版的论文集中。

照片为章程研究员向与会代表做题目为"不同土地利用土壤与岩溶作用"报告。

After the meeting, Prof. Zhang Cheng visited the karst region of San Antonio in northern Texas. Since 1971, 15 dams have been constructed in the area to control flooding and provide recharge for the Edwards Aquifer.

会后考察得克萨斯州北部圣安东尼奥(San Antonio)岩溶区、橡树公园、岩溶谷地大坝(用于拦截洪水)。自1971年以来，圣安东尼奥北部岩溶谷地区已建成15个洪水控制大坝，有效地缓解了洪涝灾害，同时还起到了增加爱德华含水层补给的作用。

International Conference on Hydrology and Disaster Management (H&DM 2009)
"水文与灾害管理"国际研讨会

The International Conference on Hydrology and Disaster Management was held in Wuhan, China from November 2 to 4, 2009. The conference was sponsored by the UNESCO Office in Jakarta, Chinese National Committee for the IHP and the Bureau of Hydrology, Ministry of Water Resources of the PRC. More than 80 representatives and experts from Australia, North Korea, South Korea, Indonesia, New Zealand, Myanmar, Philippines, Papua New Guinea, Vietnam, Malaysia, Thailand, Cambodia, Japan, Sri Lanka, Nepal, Mongolia, Laos, India, Italy and China, attended the conference.

2009年11月2-4日,"水文与灾害管理"(Hydrology and Disaster Management)国际研讨会在武汉召开。本次研讨会由联合国教科文组织雅加达办事处、国际水文计划中国国家委员会、水利部水文局主办,长江水利委员会水文局、南京水利科学研究院和国际小水电中心承办。

来自澳大利亚、朝鲜、韩国、印度尼西亚、新西兰、缅甸、菲律宾、巴布亚新几内亚、越南、马来西亚、泰国、柬埔寨、日本、斯里兰卡、尼泊尔、蒙古、老挝、瓦努哈图、印度、意大利、中国等国的IHP国家委员会代表、国内外专家学者80余人参加了研讨会。

During the conference, presentations were devoted to the focal areas of IHP-VII (2008-2013). Topics included exchange and research in the demands and experiences of Asia and the pacific and global water disaster management, sustainable development and management of water resources, and strengthening skills in these areas. Jiang Guanghui and Guo Fang from IRCK/IKG participated in the conference.

研讨会围绕国际水文计划第七阶段计划(IHP-VII 2008-2013)的5个重点领域,重点交流和研讨亚太地区乃至全球水灾害管理的需求和经验、水资源可持续发展和管理以及能力建设等问题。姜光辉和郭芳代表岩溶地质研究所和联合国教科文组织国际岩溶研究中心参加了本次研讨会。

The 6th International Climate Change – the Karst Record Conference in Birmingham, UK
参加第六届"气候变化的石笋记录"国际会议

The 6th International Climate Change–the Karst Record Conference was held at the University of Birmingham in the UK from June 26–30, 2011. A delegation from IRCK, including Prof. Yuan Daoxian, Prof. Zhang Meiliang, Dr. Qin Jungan, Dr. Guo Fang and Ms. Zhu Xiaoyan attended this conference.

Photo: Field trip to Derbyshire on June 30, 2011.

2011年6月26–30日,第六届"石笋记录的气候变化"(The 6th International Climate Change–the Karst Record Conference)国际研讨会在英国伯明翰大学召开。国际岩溶研究中心/岩溶地质研究所代表袁道先院士、张美良研究员、覃军干副研究员、郭芳副研究员、朱晓燕助理研究员等6人参加了本次会议。

照片为与会代表参加会后在Derbyshire岩溶区的野外考察。

The 4th International Symposium on Karst, Spain
参加第四届国际岩溶会议

The 4th International Symposium on Karst was held in Máaga, Spain from April 27 to 30, 2010. The symposium was organized by the Centre of Hydrogeology of the University of Máaga and the Spanish Geological Survey (IGME). More than 100 representatives from 20 countries participated, including Prof. Pei Jianguo, Prof. Zhang Cheng, Prof. Qin Xiaoqun and Dr. Guo Fang from IRCK/IKG.

2010年4月27-30日，由UNESCO、IAH岩溶专业委员会、西班牙马拉加大学和西班牙地调局共同主办，西班牙马拉加大学承办的"第四届国际岩溶会议"在马拉加大学举行，共有来自世界各地约20个国家的100多名代表参加。国际岩溶研究中心/岩溶地质研究所裴建国、章程、覃小群研究员和郭芳副研究员参加会议。

Prof. Bartolomé Andreo-Navarro from the University of Máaga explained karst in Spain.

西班牙马拉加大学Bartolomé Andreo-Navarro教授在野外讲解西班牙岩溶。

Dr. Guo Fang from IRCK gave two presentations with titles of "Problems of flood and drought in a typical peak-cluster depression karst area", and "Interpreting source of Lingshui spring by hydrogeological, chemical and isotopic methods".

Photo: Dr. Guo Fang, Dr. Lou Maurice, and Mr. Andrea Borghi were awarded the "Young Karst Researcher Prize" 2010, which was established by the IAH Karst Commission to promote karst science and encourage young people to commit to the cause of karst research.

郭芳副研究员在会上做了题为"典型岩溶峰丛洼地的旱涝问题"和"利用水文地质、水化学和同位素方法判断灵水岩溶泉的来源"2个报告。

图为IAH岩溶专业委员会主席Nico Goldscheider教授为"Young Karst Researcher Prize"奖获得者颁奖，该奖是鼓励青年学者投身岩溶研究事业、促进岩溶科学发展而设的奖项，此次会议有来自瑞士、英国和中国3位年轻人获此荣誉，郭芳是其中之一。

IRCK Attended the Scientific Workshop of UNESCO Centers in China, 2012
参加中国 UNESCO 科学工作座谈会

On January 7, 2012, Prof. Cao Jianhua, representing IRCK, attended the Scientific Workshop of UNESCO Centers in China, which was organized by the International Center on Space Technologies for Natural and Cultural Heritage (HIST) under the Auspices of UNESCO. The workshop was attended by the Executive Director of the UNESCO Natural Sciences Sector, Mr. Han Qunli, and over 20 representatives from the Chinese National Committee for the International Hydrological Programme (IHP), the Chinese National Committee for the Man and the Biosphere (MAB), the Chinese National Committee for the International Geoscience Programme (IGCP), the International Research and Training Center on Erosion and Sedimentation (IRTCES), the UNESCO Global Geopark Network Office, the UNESCO International Center on Global-scale Geochemistry (under construction) and the UNESCO Office in Beijing, attended the workshop.

2012年1月7日，国际岩溶研究中心常务副主任曹建华研究员参加了由联合国教科文组织国际自然与文化遗产空间技术中心主办的、在北京召开的"中国 UNESCO 科学工作座谈会"，参加会议的代表有来自联合国教科文组织自然科学部门执行主任韩群力先生，还有来自国际水文计划 (IHP) 中国委员会、人与生物圈 (MAB) 中国委员会、国际地球科学计划 (IGCP) 中国委员会及国际泥沙研究与培训中心、国际岩溶研究中心、国际自然与文化空间技术中心、教科文组织世界地质公园网络执行局、全球尺度地球化学国际研究中心 (筹备) 及 UNESCO 北京办事处的代表，共 20 人参加了座谈会。

Mr. Han Qunli introduced the latest advancements and trends in UNESCO Natural Scientific research. Highlighted research topics include: the water cycle, water resources and trans-boundary aquifers combined with study of global climate change; ecosystem health and biodiversity, the carbon cycle and geohazards. There is also a push to promote basic research, and application of basic research to economic and social development so as to provide proper scientific solutions to practical problems during economic growth.

会上，韩群力主任带来了教科文组织自然科学研究的最新动态和趋向，主要有两方面：(1) 强调全球气候变化下的水循环、水资源及跨边界含水层研究，生态系统健康及生物多样性研究，碳循环与地质灾害研究；(2) 强调提升基础研究、应用基础研究服务经济社会发展能力，解决经济社会发展面临的实际问题。

Photo caption:
Name: Fengshan National Geopark
Location: Northwest Guangxi
Inscribed as a National Geopark: 2010
Information: Fengshan National Geopark is known for its groupings of large caves, with huge speleothems, windows of ground rivers, and the world's highest underground sinkhole valley. Fengshan has the second largest natural bridge in China and unique karst springs.

照片说明：
名称：广西凤山地质公园
所在地点：广西西北
列入地质公园时间：2010 年
概述：作为国家地质公园，凤山是我国大型洞穴分布最为密集的地区，拥有世界大型石笋群、世界天窗群、世界最高的地下溶洞峡谷、中国跨度第二的天生桥、千古之谜鸳鸯泉等独特的地质遗迹景观。

Chapter 5

Exchange and Cooperation

第五章　交流与合作

IRCK has engaged in a wide range of exchange and cooperative activities. These including four organizational management exchanges, three academic exchanges, and six personnel exchanges. In addition, IRCK has hosted 22 academic visits for international visitors/groups, 12 cooperative memorandum (MOU) were signed, and four cooperative research projects were initiated.

国际岩溶研究中心在建设初期,开展了较为广泛的交流与合作,在交流方面,包括管理经验交流 4 批次、学术访问 3 批次、学术接待 22 批次、人员交流 6 人次,签署合作协议 12 份,开展合作研究项目 4 个。

Management Experience Exchange
管理经验交流

Visiting the Centre for Hydrogeology, University of Neuchâtel (CHYN), Switzerland
访问瑞士纳沙泰尔大学水文地质中心

At the invitation of IRCK Academic Committee member Nico Goldscheider, an IRCK delegation led by Prof. Dong Shuwen (Vice President of Chinese Academy of Geological Science) visited the Centre for Hydrogeology, University of Neuchâtel (CHYN), Switzerland from February 15 to 17, 2009. The main objective of the trip was to learn advanced scientific management approaches and management experiences from this well-established research center.

The picture shows IRCK delegation members and CHYN director Francois Zwahlen discussing areas and methods for future cooperation and signing a MOU.

在国际岩溶研究中心学术委员会委员Nico Goldscheider教授的帮助下，2009年2月15-17日，由国际岩溶研究中心理事会委员董树文教授（中国地质科学院副院长）为团长的4人国际岩溶研究中心代表团，对瑞士纳沙泰尔大学水文地质中心进行访问。此次参观考察的目的是学习该中心的科研管理模式及学科建设经验。

照片为代表团与中心主任弗朗索瓦·梓沃伦（Francois Zwahlen）教授相互介绍中心的基本情况，磋商双方合作的领域、途径，并签署双方合作意向备忘录。

During the visit, Prof. Yuan Daoxian spoke on the "Karst Dynamic System" and there was a lively discussion among researchers and students. It publicized both IRCK and new ideas and concepts of modern karstology.

访问期间，纳沙泰尔大学水文地质中心邀请袁道先院士做关于"岩溶动力系统理论"的报告，报告引起相关学者和研究生的兴趣，并热烈讨论。一方面宣传了国际岩溶研究中心；另一方面，也宣传了现代岩溶学研究的新思路、新概念。

The delegation originally planned to visit the karst landscape, caves, and springs of the Jura Mountain area. Due to heavy snow, the field trip was not able to go forth, and the visit instead focused on disciplines and lab construction. The delegation has reached two key points from the visit:

Firstly, the research team is composed of an academic area–a leader–technicians–equipments–researchers and master students or PhD students. This structure combines both science and techniques with technical staff, and thus assures data quality.

Second, the center successfully utilized microbiology research results for groundwater quality monitoring and degradation of pollutants.

访问团原定赴 Jura 山区考察岩溶地貌、燕子洞和岩溶泉的计划，因降大雪天气恶劣而被迫取消，因而，访问考察工作改为对学科建设和实验室的考察，考察后得到 2 点认识：其一，该中心的学科建设是按照学科领域 – 学科带头人 – 实验技术员 – 仪器设备 – 研究生模式组建，使得科学与技术有机结合，科学工作者技术化，使实验数据的真实性、可靠性及异常的分析落实到实处；其二，该中心成功地采用现代技术手段，将微生物相关领域的研究成果应用于对地下水质监测和水体中污染物的降解等方面。

Visit to the Institute of Water Resources Management, Hydrogeology and Geophysics (WRM), Graz, Austria
访问奥地利格拉茨水资源管理与地球物理研究所

The IRCK delegation visited the Institute of Water Resources Management, Hydrogeology and Geophysics (WRM), Austria on February 22 to 23, 2009. WRM is the largest earth science institution in Austria and cooperates widely with partners in Europe on numerous international projects. Since 1969, WRM has been funding a groundwater tracing technique training course under UNESCO. Visiting WRM was beneficial for IRCK's future international cooperation and training courses.

Photo: The IRCK delegation discussed future cooperation and signing a MOU with WRM Director Zojer Han and Prof. Ralf Benischke.

2009年2月22-23日，国际岩溶研究中心3人访问考察团对奥地利格拉茨水资源管理水文地质和地球物理研究所（WRM）进行了考察。WRM是奥地利最大的地学研究组织之一，具有广泛的欧洲合作伙伴，开展过大量的国际合作项目，1969年起承办UNESCO资助的地下水示踪技术培训班。访问WRM，对IRCK开展国际合作、国际培训班等多方面大有裨益。

照片为在国际岩溶研究中心学术委员会负责人Ralf Benischke教授的安排和陪同下，访问团与WRM研究所所长Zojer Han进行了合作意向和合作方式的讨论，并签署了合作意向备忘录。

During the meeting, the IRCK delegation visited WRM's laboratories and learned about their lab management experiences. There including isotopic lab, water chemistry lab and dye tracer lab. The good management has improved the efficiencies of the labs.

考察期间，参观了WRM实验室，并询问了实验室管理经验。该实验室从阴阳离子检测仪器的集成，不同种类同位素检测系统的转化，样品中待测指标的浓缩，玻璃器皿的清洗消毒等方面应用现代技术，使实验室的工作效率得到很大的提高。

上左图
Top left
Visiting the isotopic lab.
参观同位素监测实验室。

上右图
Top right
WRM staff explained the Failsafe data system.
参观和了解自主研发的Failsafe数据传输系统。

上图
Top
Prof. Ralf Benischke gave an introduction.
Ralf Benischke 教授在做介绍。

The IRCK delegation led by Prof. Ralf Benischke visited the underground river system in Lurgrotte Peggau cave on February 23, 2009. They observed the uses of the Failsafe data system in the field. Failsafe collects data via a dual modem connection, which is transmitted by satellite and then uploaded to internet. The main advantages of Failsafe include: (1) energy savings; (2) the ability to transmit data worldwide via the internet; (3) real-time access to data, for which any system problems can be detected and repaired promptly.

2009年2月23日，Ralf Benischke 教授陪同一起参观、考察了 Lurgrotte Peggau 洞穴，在该岩溶洞穴（地下河）系统，实地考察了 Failsafe 数据传输系统，即通过双 Modem 连接数据采集、卫星传输，数据接收，网络发布。该系统的优点主要有：（1）省力，不需很多地面建设和设施；（2）与网络连接、全球范围适用，可全球范围数据传输；（3）可最早发现系统异常并及时维修。

Visit to the Centre for Mined Land Rehabilitation (CMLR) and Isotope Lab, Sustainable Minerals Institute, University of Queensland
访问澳大利亚昆士兰大学——矿山土地修复中心和同位素实验室

Photo: Prof. Huang gave a presentation on the Sustainable Minerals Institute. Moreover, Mansour Edraki, gave a talk titled "Successful rehabilitation: a geochemical perspective", which explained mined land rehabilitation techniques and monitoring methods.

照片为黄隆斌教授就永续发展矿业研究所及矿山土地修复中心的组织、管理、研究方向、技术方法和研究成果做汇报；Mansour Edraki 做了题为 "从地球化学角度看成功的矿山治理"（Successful rehabilitation: a geochemical perspective）的报告，系统介绍了澳大利亚在矿山修复治理、矿山治理的成功技术和方法、矿山环境监测、矿山水文污染的监测。

The University of Queensland, which is one of the most famous university in Australia, was built in 1910. The university of Queensland is also a member of Group of Eight, which is Ivy League Universities of Australia.

Due to the invitation of Prof. Huang Longbin, IRCK delegation, led by director Jiang Yuchi, visited the Centre for Mined Land Rehabilitation (CMLR), Sustainable Minerals Institute, University of Queensland, Australia. A symposium with 5 presentations was held in the meeting room of the CMLR, the bilateral information was exchanged in the same time, and the potential cooperation had been discussed.

An IRCK group visited the soil lab in the Sustainable Minerals Institute. They learned that the University of Queensland offers lab management as a major for students, which is similar to library management. It provided valuable management experience for IRCK.

昆士兰大学（University of Queensland）（昆大）始建于 1910 年，是澳大利亚最大最有声望的大学之一，是被誉为"澳大利亚常青藤名校"的 Group of Eight 联盟（澳大利亚八大名校联盟）成员之一。

应澳大利亚昆士兰大学永续发展矿业研究所矿山土地修复中心 (Centre for Mined Land Rehabilitation, Sustainable Mineral Institute, University of Queensland) 黄隆斌教授（Longbin Huang）邀请，以国际岩溶研究中心主任姜玉池为团长的代表团，于 2011 年 10 月 5-12 日访问、考察了该中心。双方通过 5 个报告交流了各自的情况和关注的科学问题。

学术交流后，考察团参观了该中心的土壤实验室，同时了解到昆士兰大学将实验室管理作为一个学科专业进行教学，类似图书馆管理，这对国际岩溶研究中心 / 岩溶所实验室的管理将起到很好的指导作用。

上图
Top

Group photo, from left to right: Luo Weiqun, Li Xiaofang, Jiang Yuchi, Mansour Edraki, Huang Longbin, Cao Jianhua.
学术交流后合影留念，从左到右：罗为群、李小方、姜玉池、Mansour Edraki、黄隆斌、曹建华。

中图
Middle

Visit to the Centre for Mined Land Rehabilitation, Sustainable Mineral Institute, University of Queensland.
参观矿山土地修复中心土壤实验室。

During the 34th International Geological Congress, the IRCK representatives visited the Isotope Geochemistry Laboratory, University of Queensland. The IRCK delegation had a talk and discussed the construction of ultra-pure lab, stalagmite dating, and lab management with Prof. Zhao Jianxing and Dr. Song Yinxian.

The picture above shows Prof. Zhao explaining the isotope lab design, requirement, and function.

2012年8月，利用参加34届国际地质大会的机会，中心代表团访问了在国际上享有盛誉的昆士兰大学同位素地球化学实验室，代表团成员同实验室主任赵建新教授、宋垠先博士就超净实验室建设、石笋年龄测试及实验室管理等方面进行了交流。

照片为同位素实验室主任赵建新教授讲解同位素实验室的设计、要求和功能。

对页上左图
Oppsite page, top left
Published papers were displayed in the corridor outside the lab.
实验室外走廊中展出的研究成果。

对页上右图
Oppsite page, top right
The sample pre-preparation room where different samples are separated.
参观不同样品的前处理室的隔离、分区。

对页下左图
Oppsite page, bottom left
The IRCK delegation with Prof. Zhao Jianxin.
国际岩溶研究中心代表与实验室赵建新主任合影。

对页下右图
Oppsite page, bottom right
Air purification and filtering equipment above the lab.
实验室上隔层，空气净化的管道、过滤装置。

The Radiogenic Isotope Facility (RIF) at the University of Queensland is a HEPA-filtered ultra-clean laboratory that is widely regarded as one of the most advanced of its kind in the world.

The lab's air purification system keeps the work room at a constant temperature of 20.0±0.5 ℃ and humidity of 50%, giving it a designation of Class-10000. Water used in analysis is filtered by a Milli-Q device. To ensure chemical reagent and container purification, pure chemical reagents are distilled using a quartz distiller or double Teflon bottle distillation device. Moreover, the equipment and sample-prep rooms are divided into different sections according to their functions.

昆士兰大学同位素地球化学实验室是世界同行中一流的、现代化的开放实验室。该同位素实验室为超净实验室：空气超净，实验室常年温度保持在20.0±0.5℃，湿度保持在50%，实验室空气净化等级为Class-10000；水超净，分析用水再次经过Milli-Q装置过滤处理；化学试剂和器皿超净，分析纯化学试剂再次用石英蒸馏器或双Teflon瓶蒸馏装置蒸馏；同时仪器室、不同样品的预处理室实现功能分区。

Visit to the International Research and Training Center on Erosion and Sedimentation (IRTCES) under the Auspices of UNESCO, Beijing, China
访问国际泥沙研究培训中心

A six-member IRCK delegation led by IRCK Director Jiang Yuchi, visited the International Research and Training Center on Erosion and Sedimentation (IRTCES) under the Auspices of UNESCO on September 22, 2009. With its 25 years history, IRTCES has made noticeable achievements in its organization and coordinative functions. It has wide-ranging international influence in the field of erosion and sedimentation research and training. Periodically held international training course, founded the world Association for Sedimentation and Erosion Research, established 3 demonstrations, founded the Qian Ning Prize for Erosion and Sedimentation Technology, and started the "International Journal of Sediment Research". During the visit bilateral exchanged the construction and development, personnel trained and fostered organization management. Prof. Ning Duihu, Deputy Director of IRTCES, hosted the visit. Both sides agreed that a joint meeting should be held each year, under the sponsorship of the National Commission of the PRC for UNESCO, to share information and strengthen cooperation among the six international category II centers under the auspices of UNESCO, which are located in China.

2009年9月22日，中心主任姜玉池教授带队，国际岩溶研究中心一行6人访问联合国教科文组织国际泥沙研究培训中心。访问期间双方各自就本中心的成立与发展、人员组织结构、管理运作模式、功能与作用等问题进行了交流与座谈。

国际泥沙研究培训中心经过25年（1984-2009年）卓有成效的工作，在国际泥沙研究领域的组织与协调能力不断壮大，国际影响力不断提升，在泥沙科研与培训领域取得了令人注目的成果。主要体现：定期开展国际培训班、成立了世界泥沙研究学会（每3年举办1次国际会议）、建立了3个研究示范基地、设立了钱宁泥沙国际科学技术奖（配合世界泥沙研究学会，每3年开展1次遴选、颁奖活动）、创立《国际泥沙研究》刊物（International Journal of Sediment Research，SCI）。

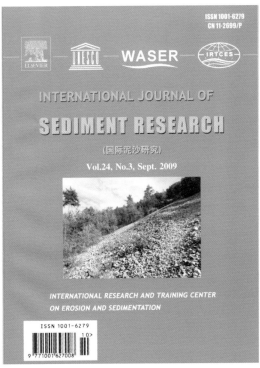

International Academic Exchange
学术交流出访

Visit to the Karst Research Institute ZRC SAZU, Slovenia
访问斯洛文尼亚岩溶研究所

From February 20 to 23, 2010, IRCK representatives (Liu Wen, Prof. Cao Jianhua, Prof. Zhang Cheng) visited the Karst Research Institute ZRC SAZU, Slovenia. It is the only European national-level karst research institute, and Slovenia is a typical and classic karst area in Europe. Even though the institute has not got a large research team, it covers almost every area of karst science, including karst topography, karst caves, karst hydrology, karst engineering, and karst biology. Although the institute is physically small, the workspace includes a meeting room, laboratory, library, editorial room and offices. The journal Acta Carsologica is also published at KRI.

2010年2月20-23日，国际岩溶研究中心考察团(Liu Wen，Cao Jianhua，Zhang Cheng)访问了该研究所。

斯洛文尼亚科学艺术研究院岩溶研究所是欧洲唯一的国立岩溶专业研究机构，斯洛文尼亚也是欧洲经典岩溶研究区。正式职工虽然不多，但研究所的研究领域几乎涵盖了岩溶学科的各个领域：岩溶地貌、岩溶洞穴、岩溶水文、岩溶工程、岩溶生物等。办公面积不大，但办公室、会议室、实验室、图书馆、编辑部有序、紧凑。该所出版的 Acta Carsologica 是国际上唯一的以岩溶命名的刊物。

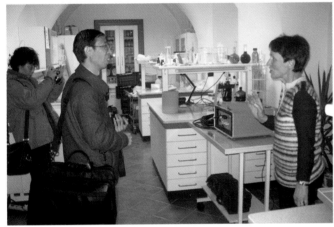

The IRCK delegation visited the KRI equipments, lab and library. Researchers from both institutes spoke and took part in academic exchange. The head of KRI, Prof. Tadej Slabe, introduced the karst cave formation model and hydrological research. At the conclusion of the visit, IRCK and KRI signed a cooperative agreement.

考察期间，代表团参观了该所的机构设置、实验室、图书馆等，与岩溶所的科研人员进行学术交流和双方基本情况介绍，所长 Tadej Slabe 还向代表介绍了岩溶洞穴形成演化的模拟及在水文地质研究方面的应用，最后签署了双方的合作意向书。

Dr. Natasa Ravbar, KRI Research Fellow, guided the IRCK group to Cerknica Polje, Planinska Jama subterranean rivers, Podstenjsek karst spring, and Skocjanske Jame UNESCO World Natural Heritage site.

在所长助理 Natasa Ravbar 博士的陪同下，访问团考察了斯洛文尼亚最大的坡立谷 (Cerknica Polje)、Planinska Jama 地下河、Podstenjsek 岩溶泉和世界自然遗产地斯科茨扬洞 (Skocjanske Jame)。

Visiting Gadjah Mada University (UGM) in Indonesia.
访问印度尼西亚卡渣玛达大学

Prof. Zhang Cheng and assistant Prof. Luo Qukan gave academic reports to the geography department. More than 50 individuals from both the school and local government participated the symposium, most of them were interested in the karst carbon process monitoring and approached the rocky desertification control and ecological rehabilitation in karst area.

章程研究员、罗劬侃博士在地理系学术厅分别做了学术报告，来自该校地理系的师生及当地政府工作人员，50多人听取了报告会，对报告中提到的岩溶碳汇监测研究、中国生态地质最新研究成果表现出了强烈的兴趣，当地政府工作人员希望能来中国实地考察。

On April 10-15, 2011, the IRCK delegation visited Gadjah Mada University (UGM), which is one of the most prestigious universities in Indonesia. UGM was established in 1949 and is located in Yogyakarta, an important cultural and educational center in Java. The island of Java is mainly comprised of late Pleistocene limestone and has well-developed karst.

2011年4月10-15日，国际岩溶研究中心访问团访问印度尼西亚（简称印尼）卡渣玛达大学（Gadjah Mada University）。卡渣玛达大学是1949年12月19日建校，是印尼三大著名学府之一，位于印尼爪哇南部的日惹，日惹是印尼重要的文化、教育中心，也是著名的爪哇文化的发源地。爪哇岛上岩溶发育，碳酸盐岩主要为晚更新世的石灰岩。

上图　Top

The delegation visited Petoyan karst spring in southern Yogyakarta with Prof. Eko Haryono. There ion concentrations in the spring water were 5.6 mmol/L HCO_3^- and 112 mg/L Ca^{2+}. In Sanglor karst spring, the ion concentrations were 6.6 mmol/L HCO_3^- and 126 mg/L Ca^{2+}.

在 Eko Haroyono 博士的陪同下，考察了日惹城南部的 Petoyan 岩溶泉，现场检测的泉水 HCO_3^- 浓度为 5.6 mmol/L、Ca^{2+} 浓度为 112 mg/L；Sanglor 岩溶泉，现场检测的泉水 HCO_3^- 浓度为 6.6 mmol/L、Ca^{2+} 浓度为 126 mg/L。

左下图　Bottom left

Prof. Suratman Wojo Suprojo from the UGM Faculty of Geography explained that the Yogyakarta region faces problems such as water loss, soil erosion, and rocky desertification. Southwest China has similar issues, during the dry season water shortage hardship for local residents.

与卡渣玛达大学地理系系主任 Suratman Wojo Suprojo 先生进行了交流，了解到日惹岩溶区面临着水土流失、石漠化的问题。中国西南岩溶区一样存在石漠化问题，旱季缺水严重，这些给当地居民的生产、生活带来很大的困难。

右下图　Bottom right

Corn plants amid exposed bedrock near Sanglor karst spring.
Sanglor 岩溶泉附近山坡的石旮旯地与生长的玉米。

Visit to the Department of Groundwater Resources (DGR), Ministry of Natural Resources and Environment of Thailand
访问泰国地下水资源厅

On April 19-22, IRCK delegation visited the Department of Groundwater Resources (DGR), Ministry of Natural and Water Resources of Thailand. With headquarter in Bangkok, DGR is an integrated department combining functions of administration and research. DGR Director General Praneet Roibang, and DGR Deputy Director General Sumrit Chusanathas were very pleased to meet the IRCK delegation. Prof. Cao Jianhua hoped to find some possibilities of cooperation on karst carbon sink, paleoclimate reconstruction, and karst aquifer research and the exploration of karst groundwater resources. Mr. Roibang stated that DGR was preparing a proposal for a comprehensive project on groundwater resources exploitation and regional hydrogeology mapping in karst region in the second half of 2011. He hoped IRCK/IKG could provide technology support and scientific guidance.

Mr. Chaiporn Sirpornpibul, Deputy Director General of the Department of Water Resources (DWR), led a field investigation to the karst landscape in

Kanchanaburi Province, northwest of Bangkok. The IRCK group also went to Chaloem Rattanakosin National Park to observe the karst topography and cave in the area. The carbonate rock in karst region, Thailand, is mainly Permian and Triassic limestone.

2011年4月19—22日，国际岩溶研究中心访问团出访泰国地下水资源厅。泰国地下水资源厅属于泰国自然资源和环境部的行政兼科研部，位于泰国首都曼谷。

代表团与泰国地下水资源厅厅长 Praneet Roibang 先生、副厅长 Sumrit Chusanathas 先生等就未来的合作进行洽谈：曹建华研究员希望能在岩溶碳汇监测研究、古气候重建研究、地下水资源开发保护等方面找到合作的机会；泰方厅长 Praneet Roibang 先生表示泰方正在准备2011年度下半年向政府申请岩溶区域地下水资源开发与区域水文地质填图项目，届时希望中方能在技术上给予指导。大家对未来合作充满期待。

随后，代表团驱车前往曼谷西北部的 Kanchanaburi 省考察泰国岩溶。在水资源厅副厅长 Chaiporn Sirpornpibul 先生陪同下，考察了国家公园（Chaloem Rat Tanakosin 国家公园）及泰国岩溶地貌、洞穴资源。泰国境内岩溶区的碳酸盐岩主要是二叠纪、三叠纪石灰岩。

Visit to the Vietnam Institute of Geosciences and Mineral Resources (VIGMR)
越南地球科学与矿产资源研究所

The Vietnam Institute of Geosciences and Mineral Resources (VIGMR), Ministry of Natural Resources and Environment of Vietnam, undertakes fundamental research on earth science, minerals and ores. From April 15 to 19, IRCK delegation visited the institute in Hanoi and a karst spring near Ha Long Bay.

During this visit, VIGMR Director Tran Tan Van gave the IRCK delegation a warm welcome. The delegation members gave three presentations to the staff of VIGMR: An introduction to IRKC/IKG; The relationship between karst environment and vegetation; and Monitoring approach for karst carbon sink. Delegation Dr. Tran Tan Van gave a report on IRCK recent advances and proposed collaborative projects on the implications of the karst carbon sink, stalagmite and paleoclimate records, and geo-parks and natural heritage sites in karst regions. The successful talk led to a MOU signed by IRCK/IKG and VIGMR.

The carbonate rock near Ha Long Bay UNESCO World Natural Heritage site is early Carboniferous limestone. The Ca^{2+} concentration is 94 mg/L and the HCO_3^- concentration is 5.1 mmol/L in the karst spring water.

越南地球科学与矿产资源研究所 (Institute of Earth Sciences and Mineral Resources of Vietnam) 隶属于越南自然资源和环境部，位于越南首都河内，是一个以基础研究为主的科研机构。2011年4月15-19日，国际岩溶研究中心访问越南地球科学与矿产资源研究所，并考察位于世界自然遗产地下龙湾保护区卡坝岛的岩溶泉。

国际岩溶研究中心委员会委员，该研究所所长 Tran Tan Van 研究员，热情接待了中心代表团。代表团在研究所就岩溶碳汇研究与监测方法、岩溶生态环境与植被相互作用等方面进行了学术交流。同时代表团 Tran Tan Van 委员汇报了国际岩溶研究中心的最近进展，并针对岩溶碳汇效应、古气候环境记录和岩溶区地质公园与自然遗产地申报等方面有意启动双方合作项目，最后，双方起草并签署了国际岩溶研究中心/中国地质科学院岩溶地质研究所与越南地球科学与矿产资源研究所谅解备忘录。

世界自然遗产地下龙湾保护区卡坝岛的碳酸盐岩为早石炭纪的石灰岩，现场检测一常年流水的岩溶泉 Ca^{2+} 浓度为 94 mg/L、HCO_3^- 浓度为 5.1 mmol/L。

Visited Western Kentucky University, University of Alabama, and Florida State University.
访问美国西肯塔基大学、阿拉巴马大学、佛罗里达大学

From August 28 to 31, an IRCK delegation visited Western Kentucky University, University of Alabama, and Florida State University.

Prof. Chris Groves of Western Kentucky University is a long-time participant in the Chinese karst research academic community. Since 1994, he has been involved in IGCP 299, 379, 448, 513 and 598. In Kentucky, the IRCK delegation visited the karst underground river system in Lost River Catchment and learned about the most recent auto-recording, auto-cleaning, and auto-calibrated field monitoring devices. They also discussed the use of stalagmites in paleoclimate research and increasing resolution by using a micro sampler and has various probes.

美国的岩溶主要分布在东部的 Appalachian 高原，即分布于佛罗里达州、田纳西州、肯塔基州、明尼苏达州、得克萨斯州、阿拉巴马州等。

2013年8月28-31日，国际岩溶研究中心访问团考察了肯塔基州的西肯塔基大学、阿拉巴马州阿拉巴马大学和佛罗里达州的佛罗里达大学。

西肯塔基大学的 Chris Groves 教授是中国岩溶学术界的老朋友，自1994年以来，他作为美国的主要组织者参加 IGCP 299、379、448、513 和 598。

在西肯塔基大学，国际岩溶研究中心代表主要考察了覆盖性的岩溶地下河系统 (Lost River)、了解了最新的美国研发的自动记录、自动清理和校正的野外监测仪器、讨论了石笋在过去气候变化研究方面的应用，以及利用微区取样器提高其分辨率。

对页上图
Oppsite page, top
Prof. Chris Groves explained the new equipment from Fondriest Environmental, Inc. and its functions. The equipment is auto-cleaning, and has various probes.
Chris Groves 教授在野外现场讲解 Fondriest Environmental 公司的最新仪器及功能，即多个电极以及电极的定时自动清理功能。

对页下图
Oppsite page, bottom
Prof. Chris Groves showing a student how to measure discharge in Lost River.
Chris Groves 教授带领学生在 Lost River 的明流段进行测流实验教学。

上图　Top
Discussing aquatic plant effects on karst carbon sink processes and methods near the exit of Lost River.
在 Lost River 出流处，讨论水生植物对岩溶碳汇迁移过程的影响及检测方法。

下图　Bottom
Discussing a stalagmite sample from Barbados (dated to 3000 years b.p.) with Prof. Jason Polk.
与 Jason Polk 博士讨论采自巴巴多斯的石笋样品（定年为 3000 年）及微区器的使用。

IRCK delegates explaining recent trends and potential publication for IRCK, and discussing the possibility of a joint journal with the editorial office of Carbonates and Evaporites.

国际岩溶研究中心代表向 James LoMoreaux 和他的同事 Ann McCarley 和 Nancy Green 介绍国际岩溶研究中心的最近的进展情况，及未来可能的出版物，同时还讨论了国际岩溶研究中心与 Carbonates and Evaporites 编辑部联合办刊的想法。

With an invitation from Dr. James LaMoreaux, who is the editor in chief of the journals Environmental Earth Sciences and Carbonates and Evaporites, IRCK delegates visited the University of Alabama from September 1 to 2. IRCK exchanged ideas with editorial offices of the two journals, visited PELA GeoEnvironmental, and discussed carbon capture and storage (CCS) and karstification in coastal area.

应阿拉巴马大学教授、Environmental Earth Sciences 和 Carbonates and Evaporites 学术刊物主编 James LoMoreaux 的邀请,国际岩溶研究中心访问团,于 2013 年 9 月 1–2 日,考察了阿拉巴马大学,与以上 2 个刊物的编辑部进行了交流,同时参观了 PELA 水文地质、环境地质、工程地质公司,并探讨二氧化碳地质储存 (CCS) 和海岸带岩溶作用过程。

Discussing CCS and coastal karstification processes with the PELA Executive Vice President Bashir Memon.

与 PELA 公司副总裁 Bashir Memon 博士探讨二氧化碳地质储存 (CCS) 和海岸带岩溶作用过程。

对页 Oppsite
Wakulla Springs and surrounding ecosystem.
Wakulla 岩溶泉的出口及周边的原始生态景观。

上图 Top
The Wakulla Springs recharge area has karst depressions and dolines.
考察 Wakulla 岩溶泉补给区及地表的岩溶洼地、漏斗。

下图 Bottom
Wakulla River and its aquatic plants.
Wakulla 地表河挺水、沉水、浮水植物。

Prof. Bill X. Hu from the Florida State University and Guy H. Means from Florida Geological Survey introduced IRCK delegates to the Florida karst and the Wakulla karst springs and river.

The discharge from Wakulla Springs is 9.5–38 m^3/s. Wakulla Spring is the headwater of the Wakulla River, the spring water flows into the Gulf of Mexico. The surrounding area has a good ecosystem.

The water of Wakulla River is clear, with abundant aquatic vegetation. It has a salinity of 140–200 mg/L, making it an excellent site for research on karst carbon sink process, especially, a great natural experimental site to study the fresh water and sea water mixing.

在佛罗里达大学 Bill Hu 教授和佛罗里达州地质调查局 Guy Means 博士的帮助下，国际岩溶研究中心代表成功地考察了佛罗里达州的岩溶和 Wakulla 岩溶泉和 Wakulla 河流。Wakulla 岩溶泉流量达到 9.5–38 m^3/s，泉水涌出后向南经 15 km 的 Wakulla 河流后汇入墨西哥湾，周边的生态系统完整且良好。Wakulla 河水清澈，有大量沉水植物、挺水植物，泉水碱度 140–200 mg/L，这些非常有利开展岩溶碳汇迁移过程的稳定性研究，尤其是陆地淡水融入海水过程的变化，是一个极好的天然试验场。

Academic Exchange
学术交流

Prof. Wolfgang Eder, Former Executive Secretary of IGCP, UNESCO Visited IRCK
教科文组织 IGCP 前秘书 Wolfgang Eder 教授来访

Prof. Wolfgang Eder, former executive secretary of IGCP, UNESCO, visited IRCK on May 25, 2009. During his tenure, Prof. Eder paid great attention to the progress and implementation of IGCP 299, 379 and 448; and gave substantial support to IRCK for its application and establishment.

Photo: Prof. Eder visited the Karst Geology Museum of China in the company of Prof. Cao Jianhua, and praised the museum as the largest professional museum for karst geology in the world.

2009年5月25日，联合国教科文组织 IGCP 前秘书 Wolfgang Eder 教授一行来到中心参观、指导。Eder 教授在任期间，一直关心与岩溶相关的 IGCP 299、379 和 448 的进展和执行情况，对国际岩溶研究中心在中国桂林建立一直给予极大的关注和支持。Eder 教授在参观了中国岩溶地质馆后，称赞该馆是国际上最大、内容齐全的岩溶地质专业科普场所。

Prof. R. Lawrence Edwards from the University of Minnesota Visited IRCK
美国明尼苏达大学 R. Lawrence Edwards 教授来访

From June 23 to 25, 2009, Prof. R. Lawrence Edwards of the University of Minnesota was invited to have a three-day academic visit to IRCK. During the visit, Prof. Yuan Daoxian introduced karst landscape of Guilin and the process of the application, preparation and establishment of IRCK. They also had a discussion about related problems of the evolution history of the Grand Canyon in USA and the Three Gorges in China and the formation of the cretaceous-tertiary red beds paleo karst. Prof. Edwards gave a lecture "The Asian Monsoon and Ice Age Terminations".

2009年6月23–25日，美国明尼苏达大学 R. Lawrence Edwards 教授应邀到中心进行学术访问。访问期间，袁道先院士向 Edwards 教授介绍了桂林的岩溶地貌和景观情况，以及国际岩溶研究中心的申报、筹备和落户等过程；Edwards 教授做了题为"亚洲季风与冰期终止点"的学术报告。

Departments of Geography and Geology of the Chinese Culture University (Taiwan) Visited IRCK
中国文化大学（台湾）地理学、地质学系师生来访

During July 15–21, 2010, 29 professors and students from the Departments of Geography and Geology of the Chinese Culture University (Taiwan) led by Prof. Kwong Fai Andrew Lo, visited IRCK as part of a cross-straits academic exchange on karst. This visit enhanced the academic exchanges between Taiwan and Mainland China, especially in the field of karst dynamic system, karst water utilization and flood and drought prevention.

2010年7月15－21日，经过中国地质科学院岩溶地质研究所和中国文化大学（台湾）的协商，在地理学系主任卢光辉教授带领下，该校地理学、地质学系师生共29人，访问了岩溶地质研究所/国际岩溶研究中心，此次访问，促进了海峡两岸的学术交流，在岩溶动力系统研究的推广、岩溶区水分利用和旱涝防治途径进行了深入探讨。

During the visit, nine academic reports were presented, three from IRCK/IKG and six from Departments of Geography and Geology of the Chinese Culture University (Taiwan). They covered recent trends in karst geology, karst geomorphology, and hydrology.

考察访问安排了学术报告会,围绕岩溶研究的最新进展和动态、地质、地貌、水文等方面,共做了9个报告,其中岩溶所/国际岩溶研究中心3个,中国文化大学6个。

左上图
Top left

Prof. Kwong Fai Andrew Lo from the Chinese Culture University (Taiwan) presented a paper on "How to face the risk of water shortage".

中国文化大学卢光辉教授做"水到用时方恨少"报告。

右上图
Top right

Prof. Yuan Daoxian participated in the seminar and explained the concept and usage of the karst dynamic system.

袁道先院士参加学术报告会,并在会上讲解岩溶动力系统。

左右下图 Bottom

Students and professors visited the Karst Geology Museum of China.

中国文化大学参观中国岩溶地质博物馆。

Field work included visit to the Maocun Karst Experimental Site and visit to Yao mountain to understand the formation mechanism of Guilin Karst Fenglin-Fengcong. Following friendly consultation and negotiation, a MOU was signed.

野外考察了桂林岩溶地貌及形成机制、毛村岩溶碳汇过程研究基地；在双方充分协商的基础上，签署了双方合作协议。

左上图 Top left
Prof. Cao Jianhua explaining karst carbon sink processes and their importance, while visiting Maocun.
考察毛村，曹建华给师生讲解岩溶碳汇过程与意义。

左下图 Bottom left
The students were very interested in karst carbon sink research and learning monitoring techniques.
同学们对岩溶碳汇研究非常感兴趣，现场学习监测技术。

Following seven days of exchanges and learning, there was a concluding meeting on the 20th. Four groups of students gave presentations of the field work and lectures.

Photo: Group photo in front of a statue of Xu Xiake.

通过 7 天的交流、野外考察和学习，20 日举行了总结汇报会，尤其是来访学生分 4 组作地质考察总结汇报，袁道先院士为学生讲解岩溶知识，师生间交流心得体会。

照片为总结会后，在中国岩溶学家徐霞客雕像前留影。

对页右上图
Opposite page, top right
At Yao Mountain, Prof. Jiang Zhongcheng explained Guilin's karst geomorphology and formation mechanism.

考察尧山，蒋忠诚讲解桂林岩溶地貌类型及成因机制。

对页右下图
Opposite page, bottom right
IRCK/IKG signed a MOU with Chinese Culture University (Taiwan).

中国文化大学与岩溶所 / 国际岩溶研究中心签署合作协议。

Barbados' First Resident Ambassador to China and the Director of Hoffman Environmental Research Institute Visited IRCK
巴巴多斯驻华大使、美国霍夫曼研究所所长来访

On September 26–30, 2010, Barbados' first resident ambassador to China, Sir Lloyd Erskine Sandiford and Prof. Chris Groves, Director of the Hoffman Environmental Research Institute, Western Kentucky University (USA) visited IRCK, and signed the Memorandum of Understanding Trilateral Cooperation on Karst Resource and Education in China, Barbados, and the United States of America. Prof. Jiang said that the MOU represents an understanding of the trilateral cooperative intentions, which can establish a solid foundation for future cooperative research, academic exchange and training activities. The international cooperation division of CAGS joined the activity.

During the visit to Guilin, Ambassador Sandiford and Prof. Groves paid a visit to the Karst

Geology Museum of China. They viewed the karst landscape along the Li River and the Longsheng terrace scenic spot, and were impressed by the typical and abundant karst resources of Guilin.

2010年9月26-30日，巴巴多斯驻华大使劳埃德·厄斯金·桑迪福德爵士，美国西肯塔基大学霍夫曼环境研究所所长克里斯·格洛夫斯博士一行访问国际岩溶研究中心，并于28日签署了《中国-巴巴多斯-美国在岩溶资源研究及教育领域的三方合作意向备忘录》（Memorandum of Understanding Trilateral Cooperation on Karst Resource Research and Education in China, Barbados, and the United States of America）。国际岩溶研究中心姜玉池主任表示，合作意向备忘录的签署体现了国际岩溶研究中心、美国西肯塔基大学霍夫曼研究所和巴巴多斯政府三方真诚合作的意向，为今后开展合作研究、学术交流和培训奠定了基础，必将极大地促进三方合作、交流和研究水平的提高。中国地质科学院国际合作处参加了接待。

在考察期间，桑迪福德爵士和格洛夫斯博士一行参观了中国岩溶地质馆、桂林岩溶地貌和龙胜梯田景观，对桂林典型的岩溶景观和丰富的岩溶资源留下了深刻印象。

A Delegation from CCOP and the Water Resources Departments of Thailand Visited IRCK/IKG
东亚东南亚地学计划协调委员会与泰国水资源厅等代表来访

From March 27 to April 1, 2011, a Thai delegation of 15 representatives from Thailand's Department of Groundwater Resources (DGR) and Department of Water Resources (DWR), Ministry of Natural Resources and Environment, the Office of Natural Resources and Environment Policy and Planning (ONEP) and the Hydrogeology Association of Thailand, led by Prof. He Qingcheng, Director of the Coordinating Committee for Geoscience Programmes in East and Southeast Asia (CCOP) Technical Secretariat, visited the IRCK and IKG.

Prof. He Qingcheng, DWR Deputy Director Chaiporn Siripornpibul, who participated in the 2010 IRCK Karst Hydrogeology and Karst Carbon Cycle Monitoring Training Course, DGR Deputy Director Supot Jermsawatdipong, and other delegates from Thailand attended the Technical Cooperation Meeting held by IRCK and IKG.

Mr. Niran Chaimanee, CCOP Geo-Environment Sector Coordinator and Mr. Chaiporn Siripornpibul introduced the basic research and management situation of CCOP, DGR and DWR, as well as the possibility of collaboration with IRCK and IKG. Prof. Yuan Daoxian, CAS Academician and Director of the IRCK Academic Committee; Prof. Dong Shuwen, Vice President of CAGS and IRCK Governing Board Member; Mr. Jiang Shijin, Director of the International Cooperation Division under the Department of Science and Technology & International Cooperation, CGS; Prof. Jiang Yuchi, Director of IRCK and IKG; and other research and management officers of IRCK and IKG attended the meeting. Moreover, the MOU of the Technical Cooperation between CCOP and IRCK was signed.

2011年3月27日–4月1日，东亚东南亚地学计划协调委员会（CCOP）秘书处主任何庆成博士及泰国自然资源与环境部水资源厅、地下水资源厅、自然资源与环境对策与计划办公室、泰国水文协会等代表一行15人访问国际岩溶研究中心/中国地质科学院岩溶地质研究所。此次来访期间，何庆成博士、泰国自然资源与环境部水资源厅副厅长Chaiporn Siripornpibul先生（他以学员的身份参加了2010年国际岩溶研究中心的国际培训班）和地下水资源厅副厅长Supot Jermsawatdipong先生分别介绍了协调委员会、水资源厅、地下水资源厅的基本情况及与国际岩溶研究中心/岩溶地质研究所的合作意向。袁道先院士、中国地质科学院副院长董树文、中国地质调查局国际合作处处长蒋仕金、中国地质科学院国际合作处处长王巍、国际岩溶研究中心/岩溶所所长主任姜玉池等参加了接待。同时，签署了东亚东南亚地学计划协调委员会–联合国教科文组织国际岩溶研究中心技术合作协议。

During the meeting, cooperation mechanisms and fields were discussed, and it was agreed to have further discussions on upcoming meetings and potential project proposals, such as international training and research projects. At the same time, an academic seminar was held, and four presentations on karst hydrology and resources; karst dynamics and global change; agricultural in karst region; and cave resources and their development, utilization and planning, were presented by sides' scientists.

访问期间，双方就合作机制与合作领域、合作内容，进行了磋商，希望在管理和技术两层面上开展密切合作，如双方联合举办国际培训活动、联合申请合作研究项目等。同时，就岩溶水文、水资源、岩溶动力系统与全球变化、岩溶区的农业、洞穴资源及开发、利用和管理，举行了学术报告会，中泰双方各做2个报告。

Top

Jiang Guanghui explained the hydrogeological structure and related monitoring data and results at Yaji Karst Experimental site.

姜光辉博士给访问团讲解丫吉试验场的水文地质结构及取得的相关成果。

Bottom

Prof. Cao Jianhua explained the karst carbon sink processes and mechanism at Maocun.

曹建华博士给访问团讲解毛村岩溶碳汇过程及机制。

The delegation visited the Yaji Karst Experimental Site and the Maocun Karst Experimental Site, learning about the geomorphology and formation of Guilin's karst.

2011年3月28-30日，访问团参观考察了桂林岩溶地貌景观和形成机制，丫吉岩溶水文试验场、毛村岩溶碳汇过程试验场。

第五章 交流与合作

上图
Top
Delegates visited a karst spring at the Yaji Karst Experimental Site, where they learned about the site's experimental construction, and the data collection processes.
访问团代表在丫吉试验场 31 号岩溶泉，详细询问观测站的建设和设计以及数据收集的过程。

下图
Bottom
Prof. Zhang Yuanhai explained the formation of the Guan Yan show cave and Guilin karst topography.
张远海博士讲解冠岩旅游洞穴的形成及桂林岩溶地貌的形成机制。

Prof. Werner E.G. Müller from Johannes Gutenberg University of Mainz, Germany Visited IRCK/IKG

德国美因茨大学 Werner E. G. Müller 教授来访

Professor Werner E.G. Müller visited IRCK/IKG from April 1 to 4, 2011 with the accompany of Prof. Dong Shuwen, Vice President of CAGS, Mr. Wang Wei, and Dr. Wang Xiaohong.

Prof. Müller gave a presentation on "Bio-Mineralization: a new research field", listened to Prof. Cao Jianhua's introduction to IRCK/IKG, and discussed the relationship between the karst environment and ecological processes.

Prof. Müller made a special visit to Prof. Yuan Daoxian, Director of the IRCK Academic Committee. They communicated and exchanged ideas on relevant scientific issues, such as the formation and development of karst dynamic system, the stability and source of bicarbonate ions in karst systems, the transform mechanisms of karst carbon sinks, and the ecological restoration and control of rocky desertification in karst regions. They also talked about the potential of cooperation in the future.

During his stay, Prof. Müller also visited the Karst Geology Museum of China and the Karst Geobiological Laboratory. IRCK/IKG's researchers' efforts and contributions to karst made a deep impression on Prof. Müller. He said he would like to promote joint cooperative work on biokarst and biogenic technologies that contribute to karst environmental rehabilitation, especially in the field of the carbonic anhydrase study. He also wished to deepen the bilateral cooperation and to provide a chance of further education for young researchers.

2011年4月1-4日，在中国地质科学院董树文副院长、国际合作处王巍处长、测试中心王晓红研究员的陪同下，国际著名生物学与古生物学家、德国美因茨大学(Mainz University)Werner E. G. Müller教授访问国际岩溶研究中心/岩溶所。

访问期间，做了题为"生物成矿新研究领域"（Bio-mineralization: a new research field）的学术报告，同时听取了曹建华研究员关于国际岩溶研究中心/岩溶所的情况介绍，并就岩溶环境与生物作用的相互关系这一课题进行了深入的讨论。

随后，Müller教授拜访了袁道先院士，两位专家就岩溶动力系统理论的形成与发展进行了深入探究，分析了碳酸氢根离子的稳定性和来源，及其对岩溶动力系统中碳的转化机理、岩溶区生态恢复及未来双方的合作领域展开了讨论。

最后，Müller教授参观了岩溶地质微生物实验室，对国际岩溶研究中心/岩溶所科研人员长期的岩溶地质工作所取得的成果留下了深刻的印象。他希望进一步推动中德双方在岩溶生态和生物技术方面的合作，更好地促进岩溶生态环境的研究和保护工作，包括对青年科技工作者的培养。

A Senior Delegation from the Federal Agency for Mineral Resources (Rosnedra), Russian Federation Visited IRCK/IKG
俄罗斯联邦地质矿产署代表团来访

A senior Russian delegation led by Dr. Anatoly Ledovskikh, head of the Federal Agency for Mineral Resources (Rosnedra), Russia, visited IRCK/IKG on October 14, 2011. CAGS Vice President Wang Xiaolie on behalf of Dr. Wang Min, Vice-Minister of China's MLR, convened the meeting. Prof. Zhang Fawang, Executive Deputy Director of IKG delivered a presentation to introduced IKG and IRCK, as well as latest research advancements.

Dr. Ledovskikh said that it was a fruitful visit and he expressed his thanks for IRCK/IKG's hospitality. He mentioned that there are karst areas in Russia, and several Russian departments are undertaking karst geology research. However, Russia has not yet an independent professional scientific research institution for karst. He hoped to enhance bilateral communication and exchange and promote karst geology research.

2011年10月14日，以Anatoly Ledovskikh署长为团长的俄罗斯联邦地质矿产署(Federal Agency of Mineral Resources (Resnedra) Russian Federation)代表团一行7人访问国际岩溶研究中心/岩溶地质研究所。中国地质科学院党委书记、副院长王小烈受汪民副部长委托主持了座谈会。岩溶所张发旺常务副所长全面介绍了岩溶所及国际岩溶研究中心的基本情况和科研地调工作进展。

Anatoly Ledovskikh署长对中心/岩溶所的热情接待和周到安排表示感谢，他说，俄罗斯也有岩溶，也有一些部门从事岩溶地质工作，希望今后能互派科研人员，加强俄中在岩溶方面的科研合作，共同推进岩溶地质研究工作。

An Official Delegation from Germany's Federal Ministry of Education and Research Visited IRCK/IKG
德国教育科研部部长一行来访

上左图
Top left
Academic exchange at IRCK/IKG.
双方在岩溶所/国际岩溶研究中心贵宾厅举办了学术交流和会谈。

上右图
Top right
The delegation visited the Karst Geology Museum of China.
访问团参观中国岩溶地质馆。

An official delegation of 10 people, led by Dr. Annette Schavan, Federal Minister of Education and Research, Germany, visited IRCK/IKG on January 11, 2012 accompanying by the officials from the Ministry of Science and Technology of the PRC, Guangxi Science and Technology Department (GXSTD), and Guilin Science and Technology Bureau. IRCK/IKG Director Jiang Yuchi and IKG Deputy Director Jiang Zhongcheng also attended the meeting.

Mr. Su Dingcheng, Deputy Director of GXSTD, gave a brief introduction on scientific development in Guangxi, focusing on the advantages of Guangxi in its good location and prospects for cooperation with countries of the Association of Southeast Asian Nations (ASEAN). Prof. Jiang Yuchi hoped this visit could help to build a cooperative bridge linking IRCK/IKG and relevant research institutions and universities in Germany. Dr. Schavan hoped to advance the exchange and communication in Clean Water Resources and Environmental Protection Cooperation between China and Germany, and to learn more about the researches advances in related fields.

2012年1月11日，德国教育科研部部长沙万女士（Annette Schavan）一行10人在中国科技部、广西科技厅、桂林市科技局相关人员的陪同下，访问了国际岩溶研究中心/岩溶地质研究所。国际岩溶研究中心主任/岩溶所所长姜玉池和岩溶所副所长蒋忠诚等参加了接待和双方会谈。

广西科技厅粟定成副厅长简要介绍了广西科技的发展概况，强调了广西与东盟各国在地理位置以及合作前途上的优势。姜玉池表示希望能通过沙万部长的此次来访，加强中心/岩溶所与德国有关科研机构、大学院所之间的合作，拓展合作领域。沙万部长表示此次来访主要是希望能在清洁水资源、维护环境问题上与中方代表进行交流和沟通，能多了解目前中方科研人员在水资源问题上、环境问题上所取得的研究进展。

上左图
Top left
IRCK Secretary-General Zhang Cheng explained IRCK's work.
章程秘书长给客人介绍国际岩溶研究中心的概况。

上右图
Top right
Visiting the MAT 253™ stable isotope mass spectrometer.
参观MAT 253稳定同位素质谱仪。

A Delegation from the Federal Institute for Geosciences and Natural Resources (BGR), Germany Visited IRCK/IKG
德国地学与自然资源研究院（BGR）

Five hydrogeology and geohazard scientists from the Federal Institute for Geosciences and Natural Resources (BGR), Germany came to IRCK during from August 22 to 26 for academic and technological exchange. This meeting was organized by IRCK/IKG, and convened by IKG Executive Deputy Director, Prof. Zhang Fawang. Prof. Thomas Himmelsbach introduced BGR and its work in groundwater investigation and protection. Prof. Jiang Zhongcheng also introduced IRCK/IKG's work in karst groundwater.

At the academic meeting, Prof. Himmelsbach presented two reports titled "BGR's activities on the groundwater sector" and "Identification of karst recharge and discharge processes using natural and artificial tracers". Dr. Georg Houben presented a report on "Groundwater protection of karst aquifers in the Near East (Lebanon, Syria)". Moreover, experts from China also presented new research results. Dr. Xia Riyuan presented a report on "Karst groundwater resources in Southwest China", and presented an initial research plan of IRCK/IKG's karst groundwater work. Dr.

下左图
Bottom left
IKG Executive Deputy Direcror Zhang Fawang welcomed the experts.
张发旺常务副所长主持接待工作。

下右图
Bottom right
Prof. Thomas Himmelsbach gave a presentation.
Thomas Himmelsbach 教授做报告。

Gan Fuping presented a report titled "Introduction of geophysical methods to detect groundwater in karst area".

The BGR representatives also visited two IRCK/IKG field research monitoring sites: the Haiyang–Zhaidi Karst Underground River System Research Site and the Yaji Karst Experimental Site. Additionally, the German experts visited the fenglin and fengcong landscape along the Li River and Reed Flute Cave in Guilin. Furthermore, they visited the Karst Dynamics Laboratory, testing center, and the karst collapse model lab at IKG.

上左图
Top left
Dr. Georg Houben gave a presentation.
Georg Houben 博士做报告。

上右图
Top right
Visit to the karst collapse modeling lab.
访问团考察岩溶塌陷模拟实验室。

2012年8月22-26日，德国地学与自然资源研究院（BGR）5名水文和地质灾害专家访问国际岩溶研究中心/岩溶所。本次合作交流会由国际岩溶研究中心/岩溶地质研究所主办，张发旺常务副所长主持会议，合作交流会上首先由德方Thomas Himmelsbach教授介绍了BGR的基本情况和在地下水调查、保护方面开展的工作；蒋忠诚研究员作国际岩溶研究中心/岩溶所情况介绍和在岩溶地下水研究方面开展的工作。

随后Thomas Himmelsbach教授和Georg Houben博士分别"通过自然和人工示踪剂的方法确定岩溶水的补给和排泄过程"（Identification of karst recharge and discharge processes using natural and artificial tracers）、"近东地区（黎巴嫩和叙利亚）的岩溶含水层保护"（Groundwater protection of karst aquifers in the Near East (Lebanon, Syria)）学术报告；中方代表夏日元、甘伏平研究员则分别做了题为"中国西南岩溶地区地下水资源"、"岩溶地区地下水的地球物理探测方法"的报告。

此次访问团还参观了岩溶动力学重点实验室、测试中心和岩溶塌陷模拟实验室，野外考察了海洋-寨底地下河观测基地和丫吉水文地质试验场。

Personal Exchange
人才交流

A graduate student from Switzerland studied at IRCK
瑞士研究生来中心学习

The Centre for Hydrogeology, University of Neuchâtel, Switzerland (CHYN) sent a graduate student, Mr. Maxime Lhotelin to IRCK for four months, starting on July 18, 2009. He studied methods for karst hydrogeological surveys, tracer tests, and the investigation of karstification intensity, and experimental analysis.

At the conclusion seminar, Mr. Lhotelin gave a presentation titled "Preferential flows in aquifer systems of the east part of Guilin, China".

2009年7月18日，瑞士纳沙泰尔大学水文地质研究中心选派了一名研究生马克西姆·洛泰林来中心开展为期4个月的岩溶水文地质结构调查、示踪试验及岩溶作用强度等方面研究工作。

调查研究结束后，马克西姆·洛泰林做了题为"中国桂林东区含水层系统中的优先流"的调研报告。

左上图 / Top left
From July 1 to August 22, 2011, IRCK Associate Researcher Guo Fang visited the Oxford Rock Breakdown Laboratory of the School of Geography and the Environment, University of Oxford, where she began her research on rock weathering in the natural environment and on man-made structures.
2011年7月1日—8月22日，郭芳副研究员访问英国牛津大学地理与环境系，开展自然环境的岩石风化和遗迹、建筑物的岩石风化学习与研究。

左下图 / Bottom left
From November 27, 2008 to May 21, 2009, Assistant Prof. Zhu Xiaoyan learned the technology of U/Th dating (238U-234U-230Th dating) for cave stalagmites at the Minnesota Isotope Laboratory of the Department of Geology and Geophysics, University of Minnesota.
2008年11月27日—2009年5月21日，朱晓燕副研究员到美国明尼苏达大学地质地球物理系同位素实验室学习石笋的U/Th定年技术。

右上图 / Top right
Under the support of the Marie Curie Action International Research Staff Exchange Scheme (IRSES), Dr. Li Qiang spent a year at Johannes Gutenberg University of Mainz (Germany) for postdoctoral research, his work focussed on the extraction, purification and classification of plant's carbonic anhydrase.
2011年10月—2012年9月，在欧盟玛丽居里人才交流项目（Marie Curie Action International Research Staff Exchange Scheme，IRSES）的资助下，李强博士于德国美因茨大学从事为期一年的访问研究，重点开展植物碳酸酐酶的提取与稳定性检测方法研究。

右下图 / Bottom right
From July to December 2012, IRCK/IKG Associate Researcher Dr. Jiang Guanghui carried out his postdoctoral work in the Department Geosciences, National Taiwan University. He concerned the hydrological network and model construction.
2012年7—12月，国际岩溶研究中心/岩溶地质研究所姜光辉副研究员在台湾大学地质科学系从事访问研究工作，重点开展水文网络与模型构建。

MOU Signing
合作协议签署

Memorandum of Cooperation Intention between the International Research Center on Karst (Institute of Karst Geology) and other organizations
(2008.12.15-2013.8)

1. Memorandum of Cooperation Intention between the International Research Center on Karst (IRCK) under the auspices of UNESCO, Guilin, China and the Center of Hydrogeology University of Neuchatel (CHYN)(February 16, 2009) **(Top left)**
2. Memorandum of Cooperation Intention between the International Research Center on Karst (IRCK) under the auspices of UNESCO, Guilin, China and the Institute of Water Resources Management, Hydrogeology and Geophysics, Joanneum Research, Graz, Austria (WRM)(February 24, 2009) **(Top right)**
3. Protocol of The Technical Cooperation Meeting between the Karst Research Institute ZRC SAZU (KRI) and the International Research Center on Karst(IRCK) under the Auspices of UNESCO (February 22, 2010)
4. Memorandum of Understanding Academic Co-operation and Exchange Program between the International Research Center on Karst(IRCK) under the Auspices of UNESCO and the Geography Faculty-the University of Gadjah Mada (FGE-UGM)(March, 2010)
5. Memorandum of Scientific Research Cooperation between the International Research Center on Karst (IRCK) under the Auspices of UNESCO and the Departments of Geography and Geology of the Chinese Culture University (Taiwan)(July 21, 2010)
6. Memorandum of Cooperation Intention between the International Research Center on Karst (IRCK) under the Auspices of UNESCO, the Government of Barbados, and the Western Kentucky University (WKU) Hoffman Environmental Research Institute(September 28,2010) **(Bottom left)**
7. Protocol of the Technical Cooperation Meeting between the Coordinating Committee for Geoscience Programmers in east and Southeast Asia (CCOP) and the International Research Center on Karst (IRCK) under the Auspices of UNESCO (March31, 2011) **(Middle right)**
8. Memorandum of Understanding between the International Research Center on Karst (IRCK) under the Auspices of UNESCO/the Institute of Karst Geology of the Chinese Academy (IKG) of Geological Sciences, Guilin China and the Vietnam Institute of Geosciences and Mineral Resources (VIGMR), Hanoi, Vietnam(April 18, 2011)
9. Letter of Intent Entered into between the UNESCO International Research Center on Karst (IRCK) and the UNESCO Chair Center in Groundwater University of the Western Cape (UWC)(December 3, 2011) **(Middle left)**
10. Memorandum of Understanding between the Center for Karst Hydrogeology, Department for Hydrogeology, Faculty of Mining and Geology, University of Belgrade (CKH) and the International Research Center on Karst (IRCK)under the Auspices of UNESCO(January 7, 2012)
11. Memorandum of Understanding between the University of Mainz and the International Research Center on Karst (IRCK) under the Auspices of UNESCO/Institute of Karst Geology (IKG), Chinese Academy of Geological Sciences(October 12, 2012)
12. Cooperation Agreement between the International Research Center on Karst (IRCK)under the Auspices of UNESCO and the Department of Groundwater Resources, Ministry of Natural Resources and Environment, Bangkok, Thailand (April 12, 2013) **(Bottom right)**

联合国教科文组织国际岩溶研究中心（中国地质科学院岩溶地质研究所）与外单位签订的合作协议

（2008年12月15日-2013年8月）

1、国际岩溶研究中心（IRCK）与纳沙泰尔大学水文地质研究中心（CHYN）合作意向备忘录（2009年2月16日）
2、国际岩溶研究中心（IRCK）与奥地利格拉兹水资源管理、水文地质与地球物理研究所（WRM）合作意向备忘录（2009年2月24日）
3、国际岩溶研究中心（IRCK）与斯洛文尼亚科学与艺术科学院科学研究中心岩溶研究所（KRI）合作意向备忘录（2010年2月22日）
4、国际岩溶研究中心（IRCK）与印度尼西亚噶速玛达大学（UGM）地理系热带岩溶研究与管理合作意向备忘录（2010年3月）
5、中国地质科学院岩溶地质研究所与中国文化大学（台湾）地质、地理学系科学研究合作协议书（2010年7月21日）
6、国际岩溶研究中心、巴巴多斯政府和西肯塔基大学霍夫曼研究所（美国）三方合作意向备忘录（2010年9月28日）
7、国际岩溶研究中心（IRCK）与东亚东南亚地学计划协调委员会（CCOP）技术合作会议纪要（2011年3月31日）
8、国际岩溶研究中心（IRCK）/岩溶地质研究所（IKG）与越南地球科学与矿产资源研究所谅解备忘录（VIGMR）(2011年4月)
9、国际岩溶研究中心（IRCK）与南非西开普大学（UWC）自然科学学院联合国教科文组织水文地质教席地下水资源管理及岩溶水文地质及边科技合作意向书（2011年12月3日）
10、国际岩溶研究中心（IRCK）与塞尔维亚贝尔格莱德大学矿产地质学院岩溶水文地质中心（CKH）合作意向备忘录（2012年1月7日）
11、国际岩溶研究中心（IRCK）与德国美因茨大学签署合作协议（2012年10月12日）
12、国际岩溶研究中心（IRCK）与泰国地下水资源厅签署合作协议（2013年4月12日）

国际岩溶研究中心 *6* 年历程

Cooperative Projects
项目合作

IGCP/SIDA Project 598
IGCP/SIDA 598 项目

Project name: IGCP/SIDA Project 598 "Environmental Change and Sustainability in Karst Systems: Relations to Climate Change and Anthropogenic Activities".

Duration: 2011–2016

Project leaders: Dr. Zhang Cheng, Prof. Chris Groves, Prof. Yuan Daoxian, Dr. Augusto Auler, Dr. Jiang Yongjun, Dr. Martin Knez, Dr. Bartolome Andreo-Navarro.

Project content:

A multi-disciplinary approach was utilized to address the four major areas of emphasis for the project that focus on key temporal and spatial scales associated with environmental change in karst systems:

(1) significantly better estimation of the carbon sink potential from carbonate rock dissolution on the continents with improvement of approaches used for these estimations that consider geobiological processes and anthropogenic influences;

(2) research on the responses of hydrogeological behaviour of karst aquifers and water resource processes under the influence of different weather and climatic events, including extreme events of droughts and floods;

(3) research on the improvement of methods for ground water vulnerability assessments to contamination and development karst disturbance indices in different karst landscape/aquifer systems;

(4) quantification of records of environmental change within water, sediments, speleothems, and cultural records preserved within karst systems that provide information over various timescales.

IGCP/SIDA 598 项目名称：岩溶系统的环境变化及可持续性——气候变化和人类活动的关系

项目执行期：2011-2016 年

项目国际工作组主席：章程（国际岩溶研究中心秘书长）

国际工作组联合主席：Augusto Auler（巴西岩溶所）、蒋勇军（西南大学地理科学学院）、Martin Knez（斯洛文尼亚岩溶所）、Bartolome Andreo-Navarro（西班牙马拉加大学水文地质中心）、袁道先（国际岩溶研究中心学术委员会主任）、Chris Groves（美国西肯塔基大学霍夫曼环境研究所）

项目研究内容：

(1) 考虑地质生物作用与人类活动影响因子，改进估算方法，更精确地估算全球碳酸盐岩溶解的碳汇潜力；

(2) 研究在不同天气和气候事件（包括旱涝极端事件）条件下的岩溶含水层和水资源过程的水文地质行为响应；

(3) 改进地下水脆弱性评价研究方法，利用岩溶扰动指数法，评估不同岩溶景观/含水层系统的脆弱性和污染现状；

(4) 提取岩溶系统中水体、沉积物、洞穴堆积物和文化记录所反映的不同时间尺度的环境变化信息，并对信息记录进行量化。

The kick-off meeting for IGCP/SIDA 598 and the International Conference on Karst Hydrogeology and Ecosystems was held at Western Kentucky University on June 8-10, 2011. The meeting was sponsored by Hoffman Environmental Research Institute, National Cave and Karst Research Institute, International Association of Hydrogeologists and Edwards Aquifer Authority. More than 60 karst scientists from China, US, Germany, Indonesia, Slovenia, Belgium, Ukraine, Hungary, Vietnam, Brazil, and Jamaica participated in the meeting.

IGCP/SIDA 598 项目的启动会，于 2011 年 6 月 8-10 日在美国西肯塔基大学举行，同时召开了"岩溶水文地质与生态系统国际研讨会"（The 2011 International Conference on Karst Hydrogeology and Ecosystems），主办方为霍夫曼环境研究所、国际洞穴岩溶研究所 (NCKRI)、国际水文地质学家协会 (IAH) 岩溶专业委员会。

来自中国、美国、德国、英国、印度尼西亚、斯洛文尼亚、比利时、乌克兰、匈牙利、越南、巴西、牙买加等多个国家 60 多名学者参加了此次会议。

上图
Top
Prof. Yuan Daoxian spoke at the meeting.
联合主席袁道先院士在会上发言。

下图
Bottom
Prof. Zhang Cheng presented a presentation at the meeting.
主席章程博士在会上做报告。

The project has been successfully carried out for two years, during which major advances were made in the areas of karst carbon sink potential, karst aquifer protection and sustainable use of water resources, and cave stalagmite and climate change record. More than 150 people participated in the project.

Figures: 2011 and 2012 annual reports for IGCP/SIDA 598.

项目已经成功地执行了 2 年，在岩溶碳汇潜力与气候变化、岩溶含水层与水资源过程、岩溶流域环境保护与水资源可持续利用、洞穴石笋对气候环境变化记录的综合解译等方面均取得可喜的进展。目前参加该项目的研究者超过 150 人。

上图是国际工作组秘书处撰写的 2011 年、2012 年的年报。

Major Advance of Project "Comparative Study of Karst Hydrogeology between China and Indo-China"
中国与中南半岛岩溶地质对比研究第一次泰国选点考察顺利开展

左上图
Top left
The Sino-Thai joint team tested spring water chemistry.
中泰项目组队员现场检测泉水水化学指标。

左下图
Bottom right
Sino-Thai joint team performing geophysics survey.
中泰项目组队员开展地球物理勘探。

Project name: "Correlation study of karst hydrogeology between China and Indo-China peninsula".
Project source: China Geological Survey
Project leader: Dr. Zhang Cheng
Duration: January 1, 2012 to December 31, 2014
Project aims are:

(1) With a research focus on the karst region in Thailand, the project was focused on comparison of the karst in China and Indo-China with regard to karst morphology, formation conditions, and development processes;

(2) Start geo-environmental investigation of the typical karstification processes and carbon cycle in Indo-China;

(3) Construct a global karst carbon sink dynamic monitoring station in Indo-China, and obtain data for estimating the global karst carbon sink;

(4) Investigate the caves and stalagmites in Indo-China, and compare the sedimentation with the climate record, and unveil the regional environmental change processes and differences under the effect of Asian monsoons.

项目名称：中国与中南半岛岩溶地质对比研究
项目来源：中国地质调查局境外地质调查项目
负 责 人：章程
起止年限：2012年1月1日-2014年12月31日
项目内容：

（1）以泰国岩溶区研究为重点，在岩溶形态、形成条件、演化规律等方面，开展中国与中南半岛岩溶研究对比；

（2）开展中南半岛典型岩溶作用与碳循环地质环境调查；

（3）建立中南半岛全球岩溶碳汇效应的动态监测站，为测算全球岩溶碳汇提供数据支持；

（4）开展中南半岛洞穴与石笋调查，进行沉积记录气候信息的对比与综合集成，揭示同为亚洲季风影响下的区域间的环境变化过程及差异。

上图
Top
Karst spring Blue Pool in Sra Morrakot, Thailand.
泰国甲米省 Sra Morrakot 岩溶泉——蓝泉（Blue Pool）。

对页上右图
Opposite page, top right
Fish Cave Eranwan in Tham Pla-Pha Sue National Park, Thailand.
泰国媚宏颂探帕拉苏国家公园（Tham Pla-Pha Sue National Park）鱼洞泉（Fish Cave Eranwan）。

KDL Open Projects
KDL 开放项目

上图
Top
Prof. Chris Groves and Prof. Zhang Cheng visited Panglong Cave in Guilin.
2011年7月，Chris Groves 和章程等考察桂林盘龙洞。

下图
Bottom
Chris Groves and Sean Vanderhoff visited Maocun in Guilin in March, 2012.
2012年3月，Chris Groves 和 Sean Vanderhoff 考察桂林毛村。

Project name: Development and Testing of a Standard Operating Procedure for a Global Network of Stations to Measure the CO_2 sink from Carbonate Mineral Weathering

Project source: KDL Open Projects

Project leader: Prof. Chris Groves, Western Kentucky University

Duration: January 1, 2012 to December 31, 2013

Project content:

(1) Standardization and testing of equipment and related methodology that will form the basis for the eventual network;

(2) Collaboration and training on karst carbon sink for US and Chinese scientists and students in the required theoretical, technical, and computational aspects of monitoring;

(3) Field data for the first time using standardized protocols will allow comparison of the CO_2 sink between two different climates (subtropical vs. subtropical monsoon) but under similar geological conditions.

项目名称：Development and Testing of a Standard Operating Procedure for a Global Network of Stations to Measure the CO_2 sink from Carbonate Mineral Weathering

项目来源：国土资源部/广西壮族自治区岩溶动力学重点实验室开放课题

负责人：美国西肯塔基大学 Chris Groves 教授

起止年限：2011年1月1日-2013年12月31日

项目内容：

(1) 监测网络的仪器设备及检测、计算方法标准研发；

(2) 举办中美联合对岩溶碳汇的概念、形成机制和监测研究方法的培训班；

(3) 通过桂林毛村地下河、Boeing Green 的 Lost River 地下河的对比研究，检验标准。

Application No.	12
Approved No.	KDL2011-03

Karst Dynamics Laboratory, MLR and GZAR

Application Form for Open Project

Title: Development and Testing of a Standard Operating Procedure for a Global Network of Stations to Measure the CO_2 sink from Carbonate Mineral Weathering

Applicant: Chris Groves, PhD

Institution: Hoffman Environmental Research Institute, Western Kentucky University

Address: 1906 College Heights Blvd. Bowling Green KY 42101 USA

Telephone: +1 270 745 5201

email: chris.groves@wku.edu

一：**Brief Message**

Description of project	Title			Procedures for a Global Network of Stations to Measure the CO_2 Sink from Carbonate Mineral Weathering			
	Duration			2012.01—2013.12	**Total budget**	100,000RMB	
Description of participants	Name	Gender	Age	Title	Assignment	Institution	Signature
	Chris Groves	m	53	Professor	Project Director	WKU	*(signed)*
	WKU staff	To be determined					
	WKU staff	To be determined					
	WKU staff	To be determined					
	IRCK staff	To be determined					
	IRCK staff	To be determined					
	IRCK staff	To be determined					

Abstract (including contents and objectives)

  The purpose of this research is to investigate and better quantify a specific process within the carbon cycle, the global sink of atmospheric carbon associated with the dissolving of carbonate minerals within limestone bedrock in the world's karst areas, regions like southwest China where soluble bedrock has been dissolved creating caves, sinkholes, and related features. When calcite and related carbonate minerals in karst

Project members from the United States visited Guilin twice, on July 8 to August 25, 2011, and March 29 to April 4, 2012, for the purposes of exchanging methodology on the karst water cycle and carbon sink process and calcification, jointly carrying out field work, and discussing monitoring data and the recommendation for standard procedure for monitory station (right page).

开发项目执行后,美方2次来桂林开展工作,分别为2011年7月8日－8月25日,2012年3月29日－4月4日,与中方开展进行岩溶水循环与碳汇监测技术方法交流,开展学术讲座和联合野外考察。

Draft recommendations for standard procedures for monitoring station to measure the carbon sink from karst process carbonate mineral weathering

Contents
1. Introduction
   1.1 Purpose and backgrounds
   1.2 Concept and definition
2. Site typical catchment selection karst
   2.1 Ratio of carbonate rock covered in KTC
   2.2 Size of KTC
   2.3 Hydrogeological structure and aquifer media
   2.4 Status of ecosystem
3. Catch many investigation and survey
4. Monitor selection
   4.1 Partial equipment for monitor
   4.2 Automatic recording data-logger for monitor
   4.3 Automatic recording and transmitting data-logger for monitor
5. Parameters requirement
   5.1 Air temperature
   5.2 Precipitation
   5.3 Discharge
   5.4 pH
   5.5 Water temperature
   5.6 Electronic conductivity
   5.7 Dissolved oxygen
   5.8 Other ions Ca^{2+}, Mg^{2+}, K^+, Na^+, HCO_3^-, Cl^-, SO_4^{2-}, NO_3^-, measured the saturation of calcite
6. Placement for data collection
   6.1 Place for data-logger size and shape, relevant steady condition
   6.2 Weir and dam for discharge
   6.3 Soil point for soil CO_2
7. Sensors and equipment recommendation
   7.1 Meter logical (air temperature, precipitation) sensors etc
   7.2 Meter quality and quantity
   7.3 Soil carbon
   7.4 Sensor and equipment protecting house
8. Monitor operation and maintenance
   8.1 Field cleaning of sensors
   8.2 Field calibration of sensors
   8.3 Substitution (replace went) sensors
   8.4 Trouble shoot
   8.5 Energy supply
   8.6 Field notes and equipment Logs

对页左上图
Opposite page, top left
The standard procedure was discussed.
Chris Groves 与中方研究人员就岩溶碳汇监测标准和相关细节进行研讨。

对页右上图
Opposite page, top right
Prof. Chris Groves discussed karst carbon sink processes and effects with Prof. Yuan Daoxian.
Groves 教授与袁道先院士研讨岩溶碳汇过程与效应。

对页左下图
Opposite page, bottom left
Prof. Chris Groves gave a presentation on "Methods for measurements of the atmospheric carbon sink from carbonate mineral weathering".
Groves 教授为大家做题为"由碳酸盐岩风化导致的大气碳汇的测量方法"的学术报告。

对页右下图
Opposite page, bottom right
(Sean Vanderhoff) from Hoffman Environmental Research Institute gave a report entitled "Delineation of groundwater basin boundaries in karst flow systems".
Sean Vanderhoff 助理研究员做"岩溶动力系统地下河流域边界确定"的学术报告。

Photo caption:
Name: Leye-Fengshan Geopark
Location: Fengshan, Baise City, Guangxi
Time to be listed as a geopark: 2010
Summary: Leye-Fengshan Geopark is located in western Guangxi, with a total area of 930 km². The park has the largest doline group, the most concentrated group of cave hall, karst window group, the natural bridge with largest span length, typical cave sediments, the most completed fossil. It has important scientific and aesthetic value.

Chapter 6
Science Dissemination and Consulting Service
第六章 科普与咨询

Since its establishment, IRCK has engaged in multiple forms of public outreach, science dissemination, and consulting service. IRCK guest-edited a special issue of "Man and Biosphere" on karst rocky desertification. IRCK also popularized advances in karst carbon sink research through scientific reports. The Karst Geology Museum of China was utilized as a base for karst scientific education. A Sino-US team conducted professional training in multiple villages and counties in Southwest China. TV programmes were made related to karst.

国际岩溶研究中心成立以来,以多种形式开展科普和咨询活动,主要包括与MAB中国国家委员会合作出版"岩溶石漠化专辑"刊物;通过科学报道,宣传中国岩溶地质碳汇研究进展;充分发挥中国岩溶地质馆的科普基地作用;中美合作深入县城和农村,开展专业培训;利用电视台制作多个岩溶科普节目。

对页图
Oppsite page

Shoot by Tian Jieyan reproduced from <Man and the Biosphere> (Issue 5, 2009)
田捷砚摄 转载自《人与生物圈》2009年第5期

Science Dissemination and Media Outreach for the "Comprehensive Control of Rocky Desertification in Karst Area" Project
为中国"岩溶区石漠化综合治理工程"项目,提供科普、咨询服务

In order to support the National engineering project "Comprehensive Control of Rocky Desertification in Karst Area", IRCK is guest editor to publish the Special Issue "Rocky Desertification", joint with the editorial of Man and Biosphere. In the special issue, popular language is used to explain the causes and origin of rocky desertification, its harmfulness, and to encourage local residents to be fully involved in the programme. The implementation of the National engineering project is in "Eleventh Five-Year Plan" and "Twelfth Five-Year Plan" of China.

Southwest China has a high population density and small amount of arable land per capita. Long-term unsustainable land use pattern has led to soil loss, rocky desertification and local people poverty.

为配合国家"十一五"、"十二五"岩溶区石漠化综合治理工程项目的实施,普及岩溶石漠化的理论知识,认识岩溶区石漠化的危害,将国家工程转化为当地百姓的自发行为,努力推进"当地百姓是治理工程的主体"的观念,国际岩溶研究中心/岩溶地质研究所与MAB中国国家委员会联合出版了石漠化专辑。

中国西南区人口密度大,人均耕地少,缺水少土,以及长期不合理的土地利用方式,导致石漠化,一方水土养活不了一方人。

第六章　科普与咨询

上图
Top

Shoot by Wu Dongjun, reprinted from <Man and the Biosphere> (Issue 5, 2009)

吴东俊摄 转载自《人与生物圈》2009 年第 5 期

Media Outreach for the "Comprehensive Control of Rocky Desertification in Karst Area" Project
国际岩溶研究中心为"岩溶区石漠化综合治理工程"项目提供咨询服务

In April 2009, Prof. Yuan was interviewed by journalists from the Chinese Economic Herald, and an article was published in the newspaper, which was titled "Control of Rocky Desertification Can Not Be Simply Equated with Afforestation". In this article, Prof. Yuan explained the causes of rocky desertification in the karst areas of southwest China. He emphasized the necessity of comprehensive control depending on local conditions, and the significance of prevention and control of underground water pollution. He also emphasized the small-watershed-unit way of management in the project (For details, please see the Chinese Economic Herald, page A2, April 15, 2009).

2009年4月,袁道先院士接受《中国经济导报》记者采访,就西南石漠化问题提出4点咨询意见:西南石漠化源自"自然条件+人口压力";因地制宜、综合治理至关重要;莫让西南13 000 km岩溶地下河变"下水道";石漠化治理要以小流域为单元。采访内容以"石漠化治理不能简单等同于植树造林"为题,刊登在4月15日《中国经济导报》上。

In June, 2009, Prof. Cao Jianhua was interviewed by journalists from the Chinese Economic Herald, and he expressed his opinion that "scientific and technological support should play a more prominent role in the project of the comprehensive control of rocky desertification". On the basis of the particularity of karst environment, the dissimilarity of space and the diversity of comprehensive control models, Prof. Cao emphasized that with regard

to the issue of comprehensive control of rocky desertification, water is essential, soil is crucial, plants are fundamental, and the harmonious development of ecology and economy leading to eradication of poverty in local populations is the objective (For details, please see the Chinese Economic Herald, page A2, June 23, 2009).

2009年6月，曹建华研究员接受《中国经济导报》记者采访，就石漠化综合治理工程中的相关问题提出5点意见：发展草食畜牧业并不适合于所有石漠化地区；石漠化因类型不同，治理模式、成本不一；科技支撑的重要性强调得还不够；多方参与，可以产生既科学又适应市场的综合治理方案；科技工作者希望参与到综合治理工程的第一线。采访内容以"科技支撑是石漠化综合治理中不可忽视的力量"为题，刊登在6月23日《中国经济导报》上。

From June 6 to June 8, 2012, Prof. Yuan Daoxian was invited to attend "The Special Symposium on Eradicating Poverty in Karst Areas of Southwest China" in Wenshanzhou, Yunnan Province. This symposium was sponsored by the state council office of poverty alleviation. More than 180 participants, including county directors and county-level Communist Party secretaries from the 80 national level poverty counties which suffer from serious rocky desertification attended the symposium. The workshop concerned the relationship between rocky desertification and poverty in karst regions, and what steps should be taken in future.

Prof. Yuan Daoxian presented a report on "Rocky desertification control in southwest karst regions of China" and talked about taking action against rocky desertification in accordance with the local conditions and paying more attention to flooding and drought. He also discussed karst groundwater contamination, and managing and utilizing scattered land and epikarst spring water in karst regions.

　　袁道先院士参加由国务院扶贫办主办"贯彻中央扶贫开发工作会议精神县级党政领导干部（滇、桂、黔石漠化区）专题研究班"。该专题研究班于 2012 年 6 月 6 – 8 日在云南文山州召开，研究班探讨了岩溶区石漠化与贫困的内在关系，及新形势下扶贫工作的重点，参加人员为来自滇、桂、黔三省岩溶石漠化区的 80 多个国家扶贫工作重点县的党政领导干部，共 180 多名。

　　袁院士在会上做了题为"我国西南岩溶地区的石漠化治理"的专题报告，并强调岩溶区石漠化治理是一个综合性的工程，需要根据岩溶发育的特点因地制宜，需要不同学科领域专业人员的相互配合，需要不同部门间的协调管理，统筹考虑生态效益、经济效益，将石漠化治理与农民的脱贫致富工作有机结合。

Chinese Research on Karst Carbon Sinks Reported in "Science"
中国岩溶碳汇效应研究进展在《科学》杂志上报道

Ms. Larson interviewed Prof. Yuan Daoxian.
采访袁道先院士。

Ms. Larson understood the karst carbon sink processes and effects at Maocun Experimental Site.
考察毛村岩溶碳汇过程与效应研究试验场。

Visit to Panlong Cave.
考察盘龙洞岩溶洞穴。

On November 18–20, 2011, Ms. Christina Larson, Beijing-based American journalist for Science, visited IRCK/IKG, with the help of Prof. Chris Groves from Western Kentucky University. Ms. Larson had a talk with IRCK Academic Committee Director Yuan Daoxian on the karst carbon sink process and mechanism. She also visited Panlong Cave and the Maocun Karst Experimental Site to understand and verify the karst carbon sink and its effects.

2011年11月18-20日，在国际岩溶研究中心理事会委员、美国西肯塔基大学教授Chris Groves的协助下，美国旅居北京的《科学》（Science）杂志新闻栏目自由撰稿人Christina Larson，访问了国际岩溶研究中心/岩溶地质研究所，Christina Larson女士在桂林期间，随中心研究人员现场考察了毛村岩溶碳汇过程与效应研究试验场、盘龙洞岩溶洞穴，并采访国际岩溶研究中心学术委员会主任袁道先院士。

Chapter 6 Science Dissemination and Consulting Service

上图
Top
The article "An Unsung Carbon Sink".
《鲜有问津的碳汇效应》（An Unsung Carbon Sink）的科学报道。

Ms. Larson interviewed several international karst experts including Dr. George Veni, director of the US National Cave and Karst Research Institute; Prof. Nico Goldscheider of the Institute of Applied Geosciences: Hydrogeology, Karlsruhe Institute of Technology, Germany; Prof. Chris Groves, director of the Hoffman Environmental Research Institute; and ecologist Li Yan from the Xinjiang Institute of Ecology and Geography, Chinese Academy of Sciences. Ms. Larson's article "An unsung carbon sink" appeared in Science vol. 334 on November 18, 2011.

　　Christina Larson 女士在收集大量第一手材料的基础上，并就岩溶作用与碳循环过程相关信息，采访了国际上多位岩溶专家，主要包括：美国国家洞穴与岩溶研究所所长 George Veni、德国 Karlsruhe 大学地质研究所水文地质学家 Nico Goldscheider、美国西肯塔基大学霍夫曼环境研究所所长 Chris Groves 和中国科学院新疆生态与地理研究所植物生态生理学家李彦。Christina Larson 女士撰写的科学新闻报道《鲜有问津的碳汇效应》（An Unsung Carbon Sink）在《科学》杂志第 334 卷（2011 年 11 月 18 日）发表。

Discussion of related scientific questions.
现场讨论相关科学问题。

Karst carbon sink monitoring and inspection in the field.
通过对比，现场考察和检测岩溶碳汇过程。

Consultation Activities of the IHP-VII (2008-2013)
积极参与IHP-VII（2008-2013）的相关咨询活动

IRCK and IAH representatives had a talk with the DIKTAS leaders.
From left to right:
Bartolomé Andreo-Navarro, Neven Kresic, Cao Jianhua, Petar Milanovic, Holger Treidel, Neno Kukuric, Zhang Cheng.

IRCK 代表、IAH 岩溶专业委员会代表与项目组织者交流，从左到右：
Bartolomé Andreo-Navarro, Neven Kresic, 曹建华, Petar Milanovic, Holger Treidel, Neno Kukuric, 章程。

Invited by Dr. Alice Aureli, representatives of IRCK Cao Jianhua and Zhang Cheng took part in the kick-off meeting of GEF-UNDP-UNESCO regional Project "Protection and Sustainable Use of the Dinaric Karst Aquifer System (DIKTAS)".

As one of the cooperative partners, IRCK will take its duty in co-organization of symposiums, training courses in relevant research fields, and offer related experience of China to make contribution to this project.

In 2011, IRCK documented a report on Sustainable Governance of Interprovincial Aquifers- a case of Yuntaishan Geopark.

应 IHP 官员 Alice Aureli 博士的邀请，国际岩溶研究中心代表曹建华、章程参加由联合国教科文组织国际水文计划（IHP）、全球环境基金（GEF）、联合国发展署（UNDP）联合资助的"第纳尔跨边界岩溶含水层保护与可持续利用"项目(DIKTAS)启动会。

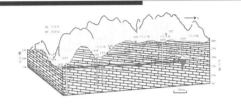

该项目于 2010 年 11 月 10-12 日在波黑特雷比涅 (Trebinje) 召开。

国际岩溶研究中心作为该项目的合作伙伴之一，通过联合举办研讨会、培训班、提供中国相关经验等方式，对该项目做出贡献。

2011 年，国际岩溶研究中心提供了《跨省界含水层可持续管理——以中国云台山地质公园为例》（Sustainable governance of interprovincial aquifers- a case of Yuntaishan Geopark）的咨询报告。

From December 3–5, 2012, IHP and the Ministry of Land and Resources have co-sponsored the consultation academic seminar: "The 4th regional consultation meeting of the Groundwater management–Global action framework for the Pan Asia Region" in Shijiazhuang city in China.

Invited by Dr. Alice Aureli, Prof. Cao Jianhua took part in the meeting and gave a presentation on Approaches to karst governance.

2012 年 12 月 3 – 5 日，由联合国教科文组织国际水文计划 (IHP) 和中国国土资源部主办，中国地质科学院水文地质环境地质研究所承办的"地下水管理——全球行动框架亚太地区第四次区域咨询研讨会"，在中国石家庄市召开。

应联合国教科文组织国际水文计划专家 Alice Aureli 博士邀请，曹建华研究员代表国际岩溶研究中心参加了会议，并做题为"岩溶地下水管理方法"（Approaches to karst governance）的报告。

左上图
Top left
Prof. Cao Jianhua represented IRCK to introduce activities and potential cooperation with IRCK DIKTAS project.
曹建华研究员代表 IRCK 介绍中心活动及与项目可能的合作。

右上图
Top right
Prof. Cao Jianhua gave a speech on the meeting in Shijiazhuang.
曹建华研究员在石家庄咨询会议上发言。

Outreach at the Karst Geology Museum of China
中国岩溶地质馆的科普基地作用

The Karst Geology Museum of China.
中国岩溶地质馆的外部景观。

Left
School students visited the museum.
中国桂林育才小学学生参观岩溶地质馆。

The Karst Geology Museum of China (KGMC) was approved as the base for karst science education on September 8, 2009. On the same day, KGMC welcomed its first visitors from primary schools and high schools.

The museum's seven exhibition halls house more than 2000 specimens, figures and models. The exhibits vividly illustrate current knowledge in karst science, and patterns and distribution of karst resources in China. The displays give an introduction on different karst fields: karst formation mechanism, karst landforms, caves and geomorphology, karst ecological systems, karst resources, karst geohazards and speleothems. The Museum is an ideal place for educational outreach about karst science.

2009年9月8日，我国第一个岩溶地质科普基地在国际岩溶研究中心正式挂牌成立，中国岩溶地质馆成为我国首批国土资源科普基地之一，并在成立当日，迎来了大批的中小学生成为首批参观者。

岩溶馆内设7个展厅，拥有展品近2000件，以大量精彩的实物标本、图片、模型生动展示了岩溶科普知识、岩溶资源状况与分布特点、我国岩溶各学科领域的概况，内容包括岩溶发育过程及其机理、岩溶地貌、洞穴景观、岩溶生态系统、岩溶资源、岩溶地质灾害、岩溶矿床等方面的基本理论与专业知识，形象地反映了我国岩溶研究现状和主要成就，展品中有大量珍贵标本，是探究岩溶科学奥秘的理想场所。

Top right

Prof. Raymond Beiersdorfer and 15 students from YSU visited IRCK. Dr. Cao Jianhua introduced Karst Dynamic System in China.

Beiersdorfer 教授于 2009 年带领 15 位学生到访国际岩溶研究中心，曹建华博士介绍中国岩溶概况。

Bottom right

Prof. Raymond Beiersdorfer and students visited the Karst Geology Museum of China, guided by Dr. Li Qiang.

国际岩溶研究中心李强博士讲解中国岩溶地质馆。

Prof. Raymond Beiersdorfer from Youngstown State University (YSU), Ohio, USA and his students visited IRCK on December 30, 2009, December 26, 2010 and January 7, 2013. The students were very interested in the Karst Geology Museum of China, and considered it as a perfect site for karst science dissemination.

在美国俄亥俄州扬斯敦州立大学雷蒙德·拜尔斯多弗（Raymond Beiersdorfer）教授的带领下，该校选修地质课程的本科生、研究生，分别于 2009 年 12 月 30 日、2010 年 12 月 26 日、2013 年 1 月 7 日，3 次参观中国岩溶地质馆。每次来访的师生对中国的岩溶现象和岩溶研究成果都兴趣盎然。中国岩溶地质馆已经成为扬斯敦州立大学到访中国必到的科普场所。

下图
Bottom
IRCK Director Jiang Yuchi with visitors from Youngstown State University.
扬斯敦州立大学师生参观后与国际岩溶研究中心主任姜玉池教授合影。

Chapter 6 Science Dissemination and Consulting Service

上图
Top
Students from Youngstown State University visited the Stalagmite Room in IRCK and asked questions about stalagmites as a record of climate change.

2010年到访的扬斯敦州立大学学生参观岩溶洞穴石笋库，并就石笋对过去气候变化的记录等问题提出询问。

左图
Left
The YSU group with the IKG/IRCK statue of Xu Xiake (the pioneer karst geographer of China, who described more than 300 caves in his Travels of Xu Xiake).

2013年到访的扬斯敦州立大学师生在徐霞客（中国最早的地理学家和旅行家，曾考察300多个岩溶洞穴，并著有《徐霞客游记》）塑像前合影留念。

March 21, 2010

Dr. Cao Jianhua
Professor
Institute of Karst Geology, CAGS, CGS
International Research Center on Karst, UNESCO
50 Qixing Road
Guilin, Guangxi, 541004
CHINA

Dear Dr. Cao,

On behalf of Youngstown State University, I express my sincere appreciation for the hospitality you extended to Professor Raymond Beiersdorfer and his students when they visited Guilin in December of 2009. The guided tour you co-lead through the Museum of Karst Geology was an important extension of the students' field trip to Shi Lin, Guilin and Yangshuo. Your lecture, which introduced the UNESCO International Research Center on Karst, the Key Laboratory of Karst Dynamics and the Institute of Karst Geology, helped the students to better understand the significance of karst as a major environmental problem in China.

The Institute of Karst Geology's location on the eastern gate of the world's broadest continuation of karstland offers students an incredible opportunity to learn about karst. The tours of the research laboratories and the stalagmite collection introduced the students to the important research being done at the research center.

To quote your former director, Dr. Yuan, I thank you for assisting Youngstown State University faculty and students in their education about "An Environment Worthy of Special Attention."

Sincerely,

Ikram Khawaja
Provost and Vice President for Academic Affairs

A thank-you note from Youngstown State University.
扬斯敦州立大学寄来的感谢函。

The Director of the Division of Science and Technology, IKG, taught the students about karst cave resources.

岩溶地质研究所科技处处长为参观者介绍岩溶洞穴及洞穴资源。

Dr. Li Qiang explained karstification, and the carbon cycle and carbon sink effect to the students.

李强博士为参观者讲述岩溶作用与碳循环过程及碳汇效应。

To celebrate the Earth Day of 2011, the Guilin Bureau of Science and Technology and Bureau of Education organized a outreach activity entitled "Securing Earth resources by adopting a new style of development", which attracted universities and high school students to visit the Karst Geology Museum of China.

2011年4月22日第42个世界地球日来临之际，中国岩溶地质馆迎来桂林市科技局、教育局举办的"珍惜地球资源 转变发展方式"科普活动的大学生、中学生代表。

115 students from the Tourism Department, and Huizhou College of Guangdong, came to Guilin to visit the Karst Geology Museum of China on May 19, 2011.

2011年5月19日，中国岩溶地质馆在"感受科学"全国科技活动周开放日迎来了广东省惠州学院旅游系地理科学专业的115名同学。

American–Chinese Cooperation for Grass–Roots Training
中美合作开展，深入基层开展培训

The Environmental Justice Young Fellows Exchange Program for American–Chinese and Scientific Dissemination on Karst Water Resources in Wuming County, Guangxi
"中美环境正义青年人才交流"项目与广西武鸣县"岩溶地下水的科学与教育"科普

The Environmental Justice Young Fellows Exchange Program was organized by Vermont Law School (VLS) and funded by the Bureau of Educational and Cultural Affairs of the US Department of State, with its aim to educate 18 active young environmental researchers (nine American, nine Chinese), especially those who come from minority groups. The project participants researched environmental problem and stated that climate change affects many minorities in the US and China. Each participant designed their own project for developing socio-environmental justice.

IRCK Associate Prof. Guo Fang took part in the Environmental Justice Young Fellows Exchange Program from May 28 to June 18, 2010.

2010年年初，郭芳经遴选，被美国佛蒙特法学院"中美环境正义青年人才交流"项目录取。该项目得到美国国务院教育与文化事务局的全额资助，目的是对18名在环境正义领域表现积极的中美青年人士(9名美国人，9名中国人)，尤其是少数族裔人士进行培训。项目参与者共同研究有关环境负担的问题，强调气候变化对中美少数族裔群体及低收入人群的影响。每个参与者将在指导下独立设计具体的项目来推动社区环境正义的发展。

2010年5月28-6月18日，郭芳在美国佛蒙特州及首都华盛顿参加了项目的交流活动。

Thanks to the Environmental Justice Young Fellows Exchange Program, in which IRCK Associate Prof. Guo Fang took part in, Ms. Leslie A. North and Prof. Jason S. Polk from the Hoffman Environmental Research Institute, Western Kentucky University were invited to attend the IRCK-organized scientific popularization and education activity in Wuming County, Guangxi, P. R. China on August 13-16, 2010.

During this course, Ms. North and Prof. Polk gave a joint educational presentation on "Protecting karst groundwater through science and education". Associate Prof. Guo Fang gave a report entitled "Karst groundwater protection and relative laws". More than 200 people participated in this training and communication workshop, including the directors and staff of the Wuming County Government, Wuming County People's Political Consultative Committee, Water Resource Bureau, Environmental Protection Bureau, Forestry Bureau, Land and Resources Bureau, Tourism Bureau and the Wuming Water Supply Company, local villagers and graduate students from Southwest University of China, China University of Geosciences (Wuhan) and Guilin University of Technology and Guangxi Normal University.

2010年8月13-16日，郭芳副研究员借助该项目，邀请美国西肯塔基大学霍夫曼环境研究所的Leslie A. North女士和Jason S. Polk博士到中国广西少数民族地区武鸣县开展保护岩溶水知识的教育培训。

Leslie A. North女士和Jason S. Polk博士共同做了题为"科学教育与岩溶地下水保护"((Protecting Karst Groundwater through Science and Education)的报告；郭芳做了题为"岩溶地下水保护及相关法律"(Karst Groundwater Protection and Relative Laws)的报告。来自武鸣县政府、县政协、水利局、环保局、旅游局、林业局、国土局、供水总公司、西南大学、中国地质大学(武汉)、桂林理工科技大学的研究生等200多人到会。

对页下图
Oppsite page, bottom

The term of Environmental Justice Young Fellows Exchange Program in VLS.

"中美环境正义青年人才交流"项目参与者，在美国佛蒙特法学院。

"Major Environmental Geological Problems and Countermeasures of Karst Mountain Region in Southwest China" Seminar
"中国西南岩溶石山地区重大环境地质问题及对策研究"咨询研讨会

The "Karst Resource and Environment" Training and "Major Environmental Geological Problems and Countermeasures of Karst Mountain Areas in Southwest China" Seminar were jointly held by Western Kentucky University (WKU), IRCK and IKG in Kunming, Yunnan Province, China, from August 11 to 13, 2009. Experts from IRCK, IKG, WKU, the Karst Research Institute of Slovenia and other Chinese and American institutes and universities (including Mammoth Cave National Park and Southwest University of China) took part in the meeting.

After the training course, a seminar focusing on human activities, especially karst groundwater environment problems that caused by mining was held. The following suggestions were put forward:

(1) To strengthen further investigation to prepare a classification of major geological environment problems of southwest karst region;

(2) To intensify outreach efforts to raise awareness of karst groundwater vulnerability and the structural characteristics of the karst ecological environment;

(3) To learn the latest ideas, measures and techniques for karst groundwater resource protection from US and other western countries, so as to bring in new technologies, methods and ideas to enhance geological investigation and scientific research.

2009年8月11-13日，由美国西肯塔基大学、国际岩溶研究中心/岩溶地质研究所共同举办的"中国西南岩溶石山地区重大环境地质问题及对策研究咨询研讨会"及"岩溶地区资源管理方法与准则培训班"，于云南省昆明市举行。来自国际岩溶研究中心、岩溶所、美国西肯塔基大学和斯洛文尼亚岩溶研究所、美国猛犸洞国家公园、西南大学等98人参加了培训班。

培训班结束后，随即召开了西南矿山与水环境重大环境地质问题咨询研讨会。着重研讨人类工程活动特别是矿山开采给岩溶区地下水带来的水环境问题，并就这些问题提出了以下咨询建议：①进一步加强调查，尽快编制一张西南岩溶区重大环境地质问题的分类图，这一工作列入次年工作方案；②强化对岩溶生态环境结构特征、岩溶地下水脆弱性的宣传力度，提高广大居民对岩溶水环境的保护意识；③与美国及其他西方国家对岩溶地下水资源保护的观念、保护措施、管理机制还存在较大的差距，在加强地质调查与科研相结合的基础上，尽快引进新技术、新方法。

Media Outreach on CCTV 10 – Science and Education Channel to Popularize Karst Ecological Environment Protection
与 CCTV10 科教频道合作，制作岩溶生态环境保护的宣传节目

From January 8 to 14, 2011, Prof. Cao Jianhua, IRCK Executive Deputy Director; Dr. He Shiyi, IKG Senior Researcher; Mr. Wu Xia, IRCK Assistant Researcher; and three graduate students conducted a field investigation at the Maolan National Natural Reserve in Libo, Guizhou Province. CCTV 10's GeoChina Programme filmed their field work. The footage was broadcasted on CCTV 10 Science and Education Channel on March 13, 2011.

During this investigation, the following five questions were answered:

(1) What is the difference between the Maolan Karst Forest and other karst landforms?

(2) How does the vegetation in Maolan differ from that of the surrounding area?

(3) What is the cause of plant diversity in the Maolan Karst Forest?

(4) How did the plants in Maolan survive and evolve? What is their impact on the karst landforms of the forest?

(5) How does Maolan Karst Forest contribute to rocky desertification control?

2011年1月8－14日，何师意研究员带领国际岩溶研究中心/岩溶所青年科研人员和学生赴贵州茂兰自然保护区开展野外考察。中央电视台CCTV10科教频道《地理·中国》栏目组对此次地质考察活动进行了全程跟踪拍摄。考察的结果以"不寻常的森林"为题于2011年3月13日播出。

此次考察回答了5个问题：①茂兰岩溶森林地区与其他岩溶地貌有何不同？②茂兰森林区与其周边地区植被种类有何差异？③为何在茂兰岩溶森林地区能够形成如此丰富的植被？④此地植被以何种方式影响并改变着此地岩溶地貌？⑤石漠化对此地森林植被的影响？如何防止石漠化？

上图
Oppsite page, top left
Preparations before the fieldwork.
拍摄前的试镜。

IRCK Took Some Active Part in Science Popularization and Outreach Activities
积极参与其他各类科普活动

IRCK Academic Committee Director Yuan Daoxian attended the Kick-off Meeting of "My Low-carbon Life: Young People's Scientific Investigation Experience Activity" in Guizhou Province on June 20, 2010. At the meeting he delivered a report on karst processes, the carbon cycle and global climate change, in order to raise awareness of karst science and the environment protection of middle school students, in Guiyang city. He answered questions raised by the students, covering topics including the impacts of acid rain on the carbon sink, the source of carbon, the ocean as a carbon sink and the effects of human disturbance on the karst carbon sink.

As a result of the meeting, the young people realized not only the importance

of low-carbon living, but also the significant role of scientific research in the addressing global climate change.

2010年6月20日,国际岩溶研究中心学术委员会主任袁道先院士参加由贵州省科学技术协会、贵州省教育厅、贵州省国土资源厅及贵州省科技厅举办的"我的低碳生活——青少年科学调查体验活动"启动仪式。

启动仪式上袁道先院士为贵阳市高中生做了"岩溶作用与碳循环及全球气候变化"的科普报告,并回答了酸雨对岩溶碳汇的影响、海洋中碳源与碳汇过程与功能,以及如何利用岩溶进行碳汇回收的人为干扰等问题。

本次活动不仅让青少年朋友认识到低碳生活的重要性,还通过普及岩溶作用在全球变化中的重要作用,让大家认识到科学研究在全球气候变化中的重要地位。

Prof. Yuan Daoxian was invited to attend The Eco-Forum Global, Guiyang (EFG-2012) "Sustainable Water Resources Management" on July 26, 2012.

More than 1200 scientists and officials, including many international experts, attended EFG-2012, with the goal of seeking a sustainable approach on synchronous development for ecological civilization and industrialization. The forum was convened by Mr. Hans d'Orville, Assistant Director-General for Strategic Planning, UNESCO.

Prof. Yuan Daoxian pointed out that global extreme climate events had highlighted the importance of making rational use of water resources. For rational exploitation and utilization of water resources, it is very important to study the karst water resources around the world, especially in the typical karst areas of Southwest China. Meanwhile, we need to combine scientific research, science popularization, law enforcement and technical support.

2012年7月26日，国际岩溶研究中心学术委员会主任袁道先院士参加了第四次生态文明贵阳年会，此次年会的主题是"可持续水资源管理"（Sustainable Water Resources Management）。

此次会议共有国家领导人、国外政要、中央部委领导、国际组织代表、国内外城市负责人、中外大学校长、中外企业家、专家学者约1200人参加。会议特邀联合国教科文组织战略规划助理总干事汉斯·道维勒(Hans d'Orville)任嘉宾主持。

袁道先院士在会上指出：全球范围内的极端气候引发人们对水资源合理利用重要性的思考。为做到水资源合理开发和利用，研究世界岩溶水尤其是西南地区岩溶水具有极其重要的意义，要把科学研究、科普、执法和技术研究结合起来。

To celebrate the 2011 World Water Day and China's Water Week, IRCK took "Challenges of Karst Water Resources" as a theme for a series of educational and publicity activities concerning karst-related water environment science and water conservation on March 21 and 22, 2011, in conjunction with IKG, KDL, the Commission on Karst Geology of the GSC, Guangxi Normal University (GXNU) and Guilin University of Technology (GLUT). IRCK presented scientific posters and gave popular science lectures to fully illustrate the significance, domestic and international status, and future trends of karst water resource protection, to raise the public awareness of the karst water cycle and resource protection among young people, and to appeal to everyone to take action to protect and save water – one of the most important resources on Earth for human survival.

In addition, 12 scientific posters were produced and four lectures were given on the theme of "Challenges of Karst Water Resources" at the College of Life Sciences, GXNU and the College of Environmental Science and Engineering, GLUT. The poster topics included: ① aims of the Water Day and Water Week's activities; ② origin and significance of World Water Day;

③ the global water cycle; ④ water resource utilization and related environmental problems; ⑤ global water pollution; ⑥ the karst water cycle; ⑦ water pollution in karst regions; ⑧ drought in Southwest China; ⑨ flooding in Southwest China; ⑩ water conservation; ⑪ water laws and regulations; and ⑫ protection and sustainable use of the Dinaric karst trans-boundary aquifer system (provided by the DIKTAS project group).

2011年3月21—22日，国际岩溶研究中心联合岩溶地质研究所、国土资源部岩溶动力学重点实验室、中国地质学会岩溶地质专业委员会、广西师范大学、桂林理工大学举办2011年"世界水日"、"中国水周"宣传教育活动。以科普展览、环保讲座进校园等方式倡导保护岩溶水资源，号召大家行动起来，共同保护水这一地球上最珍贵的生存资源。

中心围绕"岩溶水资源面临挑战"这一活动主题制作了12块宣传展板，在广西师范大学、桂林理工大学举办科普学术讲座4场，营造出浓厚的宣传教育氛围。12块展板的主题：世界水日活动宗旨目的、世界水日的由来和意义、全球水循环、水资源与环境地质灾害、全球水污染现状、岩溶地下水循环、岩溶地下水污染、中国西南岩溶区干旱问题、中国西南岩溶区内涝问题、节约用水、相关法规和第纳尔跨边界岩溶含水层保护和可持续利用。

对页图
Oppsite page
Panel provided by the DIKTAS project group on protection and sustainable use of the Dinaric karst trans-boundary aquifer system.
由DIKTAS项目提供的展板"第纳尔跨边界岩溶含水层保护和可持续利用"。

左上图
Top left
The Secretary-General of IRCK, Prof. Zhang Cheng gave a report.
国际岩溶研究中心秘书长章程博士做科普报告。

右上图
Top right
Students of the Guangxi Normal University studied the panels.
广西师范大学学生观看宣传展板。

Photo caption:
Name: Shuanghe Cave National Geopark
Location: Suiyang, Zunyi City, Guizhou Province
Time in geopark: 2004
Summary: Shuanghe Cave National Geopark is located on Dalou Shan mountain, at elevations ranging from 600 to 1700 m. The park includes an array of karst geomorphic forms, including several types of fengcong karst, blind valleys, skylights, underground rivers and shafts.

照片说明：
名称：绥阳双河洞国家地质公园
所在地点：贵州省遵义市绥阳县
列入地质公园时间：2004 年
概述：双河洞国家地质公园属于大娄山山脉，海拔 600—1700 m，地形切割强烈，相对高差大，地貌类型除太阳山、金林山一线至干河沟为构造侵蚀中山外，多数地区为喀斯特峰丛洼地及峰丛谷地，形成了溶洞、峰丛谷地、峰丛洼地、盲谷、天窗、地下河、竖井、天坑等地貌形态。

Chapter 7

International Training

第七章　国际培训

Since 2009, IRCK has hosted an annual international training course. The sixth will be held in October, 2014.

In total 87 trainees from 24 countries participated in the five training courses, and 65 famous karst experts from 17 countries served as lecturers. The training course topics were: Karst Hydrogeology and Karst Ecosystem (2009), Karst Hydrogeology and Karst Carbon Cycle Monitoring (2010), Karst Hydrogeological Investigation Technology and Methodology (2011), Karst and Hydrogeochemistry (2012), Karst Hydrogeological Survey, Dynamic Monitoring and Application in River Basins(2013).

Some trainees participated in IRCK's training courses twice. When they got back to their own countries, some trainees began the Karst Dynamic System study.

国际岩溶研究中心自2009年以来，每年举办一期国际培训班，已经成功举办5次培训班，第六次培训班将在2014年10月份举行。

前5次培训班共吸引了24个国家87位学员参加，邀请了17个国家65位著名的岩溶学家作为培训班的教员，5次培训班的主题分别为：岩溶水文地质与生态（2009）、岩溶水文地质与岩溶碳循环（2010）、岩溶水文地质调查技术方法（2011）、岩溶与水文地球化学（2012）、流域岩溶水文地质调查、动态监测与应用（2013）。

很多国家的学员结束培训班的学习回国后，再次回访，并对合作和开展本国的岩溶研究做了大量的工作。

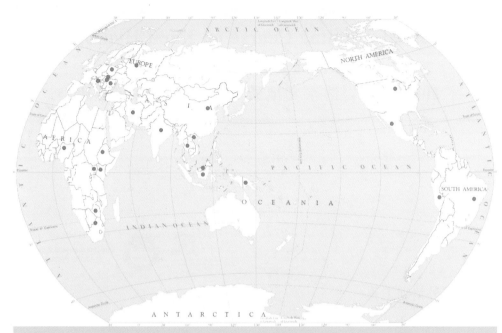

Nationalities of the students for the five training courses
5次国际培训班学员来自的国家

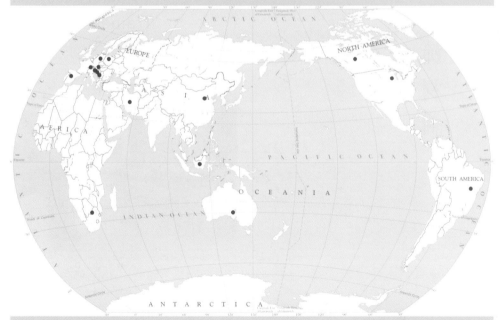

Nationalities of the lecturers for the five training courses
5次国际培训班教员来自的国家

International Training Course on Karst Hydrogeology and Karst Ecosystem, 2009
2009年"岩溶水文地质与生态"国际培训班

The International Training Course on Karst Hydrogeology and Karst Ecosystem was sponsored by the Ministry of Commerce of China and organized jointly by IRCK and IKG. It was held from November 8 to December 5, 2009 in Guilin, China, 17 trainees from Ethiopia, India, Indonesia, Kenya, Peru, Romania, Uganda and Vietnam, attended the training course.

The training course in 2009 was the first such course organized by IRCK. With the help of the Ministry of Commerce of China, the staff of the training course learned the related training requirements and provisions by the work experience exchange meeting.

To ensure the success of the training course, IRCK established a training work leading group, secretariat group and meeting group, and prepared the training materials in advance.

There were six main components for the course: an opening ceremony, lectures, practical fieldwork, field excursions, a final evaluation test of the trainees and a closing ceremony.

The lectures during the training course covered a wide range of topics including karst dynamics, karst hydrogeology and the karst ecosystem.

由中国商务部主办，国际岩溶研究中心与岩溶所联合承办的"岩溶水文地质与生态"国际培训班于2009年11月8日至12月5日在广西桂林召开。共招收学员17人，他们分别来自埃塞俄比亚、印度、印度尼西亚、肯尼亚、秘鲁、罗马尼亚、乌干达和越南8个国家。

此次国际培训班是国际岩溶研究中心组织的第一次培训活动，在中国商务部的帮助下，参与培训班的骨干，参加了商务部援外培训工作的经验交流会，学习了商务部援外培训要求的相关规定。为了做好培训，中心还成立了培训工作领导小组、秘书组和会议组，并提前准备了相关的培训材料。

根据商务部要求，此次培训包括开幕式、课堂讲学、野外实践、野外考察、学员评估和闭幕式共六个部分。

授课内容覆盖面广，基本涵盖了岩溶动力学、岩溶水文地质学和岩溶生态系统三个方面的内容。

The opening ceremony was held in the IKG auditorium on November 16. The training course was designed not only to share karst knowledge and techniques with developing countries, but also to expand exchanges and cooperation, and enhance friendships in these areas. Several distinguished guests spoke at the opening ceremony: Mr. Sun Baoliang, Deputy Director of the Department of International Cooperation & Science and Technology, MLR; Mr. Jiang Shijin, Director of the International Cooperation Division of the Department of Science and Technology & International Cooperation, CGS; Prof. Dong Shuwen, Vice-President of the Chinese Academy of Geological Sciences (CAGS); Mr. Wang Wei, Director of the International Cooperation Department, CAGS; Mr. Zhang Xiaofei, Director of the International Cooperation Department of the Science and Technology Agency of Guangxi; IRCK/IKG Director Jiang Yuchi and IRCK Academic Committee Director Yuan Daoxian.

Photo: In the front row from right to left: Jiang Zhongcheng, Wang Wei, Jiang Yuchi, Jiang Shijin, Yuan Daoxian, Sun Baoliang, Dong Shuwen, Nataša Ravbar, Zhang Xiaofei, Cozma Carnnen Madalina (trainee from Romania), Qi Shihua, Daniella Ivonne Cardenas Palma.

2009年11月16日下午16：00-17：30于岩溶所大礼堂举行开班仪式，国土资源部科技与国际合作司副司长孙宝亮、中国地质调查局科技外事部外事处处长蒋仕金、中国地质科学院副院长董树文、中国地质科学院国际合作处处长王巍、广西壮族自治区科技厅国际合作处处长张晓飞、国际岩溶研究中心主任、岩溶所所长姜玉池、国际岩溶研究中心学术委员会主任袁道先分别致辞。

此次培训班的开班意味着国际岩溶研究中心在全球岩溶地质科学领域，尤其是在发展中国家中正发挥着积极传播岩溶知识、扩大交流合作、增进友谊、广交五洲朋友的作用。

前排从右到左：蒋忠诚、王巍、姜玉池、蒋仕金、袁道先、孙宝亮、董树文、Nataša Ravbar、张晓飞、Cozma Carnnen Madalina（罗马尼亚学员）、祁士华、Daniella Ivonne Cardenas Palma（秘鲁学员）。

Chapter 7 International Training

IRCK: The First *6* Years

The training course included a total of 48 hours of classroom lectures taught by 11 karst geology experts from around the world (Austria, Australia, Canada, Croatia, Germany, Poland, Serbia, Slovenia, and the United States) and 14 domestic experts.

1. Prof. Andrzej Tyc lectured on "Karst environment, karst processes and carbon cycle in Poland".
2. Dr. Lian Yanqing lectured on "Karst hydrology and modeling principles".
3. Prof. Wang Mingzhang lectured on "Guizhou typical underground river water resources development and utilization case analysis".
4. Prof. Yuan Daoxian lectured on "Origin, structure and function of the karst dynamic system".
5. Prof. Wang Shijie lectured on "Carbonate rock weathering and limestone soil formation".
6. Prof. Derek Ford lectured on "Pollution and protection in karst aquifers: three case studies from the Niagaran Dolomites of Ontario, Canada".
7. Dr. Wilhelm Struckmeier lectured on "Global hydrogeological map and water management".
8. Dr. Nataša Ravbar lectured on "Karst in Slovenia with special regard to hydrological systems".
9. Prof. Elery Hamilton-Smith lectured on "Karst heritage: looking back and looking forward".

25名国内外岩溶地质领域的著名专家学者应邀加入培训班讲师团，为学员们授课讲学。其中，国内专家14名，外籍专家11名，分别来自美国、奥地利、澳大利亚、加拿大、克罗地亚、德国、波兰、塞尔维亚和斯洛文尼亚共9个国家，授课时间48小时。

1. Andrzej Tyc：波兰岩溶环境、过程及碳循环。
2. 连炎清：岩溶水文及建模原理。
3. 王明章：贵州典型地下河水资源开发利用案例分析。
4. 袁道先：岩溶动力系统的起源、结构和功能。
5. 王世杰：碳酸盐岩风化和石灰土的形成。
6. Derek Ford：加拿大安大略地区岩溶含水层的污染和保护。
7. Wilhelm Struckmeier：全球水文地质图与水资源管理。
8. Nataša Ravbar：斯洛文尼亚地区重点关注水文特点下的岩溶。
9. Elery Hamilton-Smith：岩溶遗产的回顾与展望。

右上图
Top right

Prof. Ralf Benischke explained tracing methods at Maocun.
Ralf Benischke教授带队在毛村地下河讲授示踪方法。

右中图
Middle right

Prof. He Shiyi demonstrated the portable field equipment for hydrochemistry monitoring.
何师意演示野外便携式仪器对水化学监测方法。

右下图
Bottom right

Dr. Jiang Guanghui showed trainees the groundwater hydrochemistry auto monitoring equipment at Yaji.
姜光辉带队考察丫吉，并演示地下河水化学自动化监测仪器的工作原理。

There were three days field work. One day was spent at Yaji Karst Experimental Site studying groundwater stream hydrochemistry auto monitoring equipment and principles, and participants practiced operating the equipment in the field. Two days were spent at Maocun Experimental Site focusing on procedures and methods for groundwater tracing.

培训班安排了3天的野外实习，1天到桂林丫吉岩溶试验场，讲授地下河水化学自动化监测仪器及原理，并对野外仪器技术的掌握及数据的获得进行实地操作；剩余2天到毛村讲授野外地下河示踪试验的进行操作规程和方法，掌握便携式水质监测技术方法。

左上、左中、左下图
Top left, middle left and bottom left.

Trainee practiced in the field.
学员们对野外监测工作兴趣盎然，争相动手一试。

Field trips to the karst landscapes in Guilin, Guangxi and the Stone Forest in Yunnan, gave the trainees more understanding of the karst dynamic system and its function, and also let they know karst environment vulnerability, the detrimental effects of rocky desertification, and Chinese Minority Romantic Feelings.

为了配合培训班的内容，中心还安排了广西桂林、云南石林岩溶景观的野外参观。这不仅从学术上让学员了解岩溶动力系统及其功能；同时也了解岩溶区的脆弱性、石漠化的危害性，了解中国少数民族风情。

对页上左图
Oppsite page, top left
Trainees and lecturers visited Seven Star Cave in Guilin, and shared the celebration of the 60th anniversary of the foundation of China.
培训班教员、学员参观桂林七星岩,并共庆中国成立60周年。

对页上右图
Oppsite page, top right
The training course should included Chinese Kungfu in the programe.
培训班真应该有"中国功夫"课程。

上左图
Top left
Trainees and lecturers visited the Stone Forest in Yunnan.
培训班教员、学员参观云南石林。

上右图
Top right
We became the Xiao Erhei to be the dream lover of Chinese AShima.
咱也当一回"小二黑",做一回中国阿诗玛的"梦中情人"。

Mr. Joseph Nganga Kuria from Kenya gave an assessment report.

肯尼亚学员 Joseph Nganga KURIA 做评估报告。

Mr. Galin Bogdan Cengher from Romania gave an assessment report.

罗马尼亚学员 Galin Bogdan CENGHER 做评估报告。

The international training course promoted communication between the trainers and trainees, and to better understand the situation of the trainees, assessments were held. The trainees gave reports on the karst landscape, karst resources and environmental problems of their own countries, and plans for future work.

国际培训班是一个很好的交流平台,为了更好地促进教员、学员深入交流,了解学员的学习情况,培训班还专门安排了学员评估环节,评估报告需要学员介绍本国的岩溶景观、岩溶资源、环境问题,以及通过学习对未来工作的设想。

The instructors gave a summary of the assessments.
评估讲师对评估结果做总结。

上左图
Top left

IGCP Executive Secretary and member of IRCK Governing Board, Dr. Robert Missotten gave a speech at the closing ceremony.

IGCP 秘书长、国际岩溶研究中心理事会委员 Robert Missotten 博士在结业典礼上致辞。

上右图
Top right

The trainees were awarded certificates.

嘉宾为学员们颁发培训班结业证。

The IRCK Governing Board and Academic Committee meetings were held immediately after the training course closing ceremony on December 5. This provided a precious opportunity for many karst experts and young researchers to interact in Guilin. A number of distinguished guests attended the event, including: Prof. Jiang Jianjun, General Director of the Department of International Cooperation & Science and Technology, MLR; Ma Yongzheng, Director of the Foreign Affairs Department of International Cooperation & Science and Technology, MLR; Mr. Jiang Shijin, Director of the International Cooperation Division of the Department of Science and Technology & International Cooperation, CGS; Dr. Dong Shuwen, Vice President of the Chinese Academy of Geological Sciences; Mr. Wang Wei, Director of International Cooperation department; members of the IRCK Governing Board and Academic Committee; and IKG leaders.

12月5日，培训班结束，赶上国际岩溶研究中心理事会、学术委员会召开，中外岩溶专家、年轻后生力量聚集桂林，是一次难得的机遇。结业典礼在岩溶所大礼堂举行，参加的嘉宾有国土资源部科技与国际合作司司长姜建军、科技与国际合作司外事处处长马永正、中国地质调查局科技外事部外事处处长蒋仕金、中国地质科学院副院长董树文、国际合作处处长王巍、国际岩溶研究中心理事会、学术委员会成员及中国地质科学院岩溶地质研究所领导。

Wonderful international party on karst:

First row (left to right): Jiang Shijin, Lu Yaoru, Wilhelm Struckmeier, Petar Milanovic, Jiang Jianjun, Derek Ford, Zhang Hongren, Robert Missotten, Yuan Daoxian, Chris Groves, Wang Jiyang, Dong Shuwen, Jiang Yuchi.

Second row (left to right): Jiang Zhongcheng, Cao Jianhua, Larry Edwards, Fredrick Siewers, Liu Wen, Huang Qingda, Wang Wei, Ma Yongzheng, Andrzej Tyc, Neven Kresic, He Shiyi, Franscois Zwahlen, Xie Yunqiu, Fei Yue.

Third row(left four): Andrej Kranjc.

国际岩溶大家庭的一次聚会：
第一排（左到右）：蒋仕金、卢耀如、Wilhelm Struckmeier、Petar Milanovic、姜建军、Derek Ford、张宏仁、Robert Missotten、袁道先、Chris Groves、汪集旸、董树文、姜玉池。
第二排（左到右）：蒋忠诚、曹建华、Larry Edwards、Fredrick Siewers、刘文、黄庆达、王巍、马永正、Andrzej Tyc、Neven Kresic、何师意、Franscois Zwahlen、谢运球、费玥。
第三排左4为 Andrej Kanjc。

International Training Course on Karst Hydrogeology and Karst Carbon Cycle Monitoring, 2010
2010年"岩溶水文地质与岩溶碳循环监测"国际培训班

The International Training Course on "Karst Hydrogeology and Karst Carbon Cycle Monitoring" was jointly organized by IRCK and IKG from November 29 to December 10, 2010. 17 course participants from 11 countries (Brazil, China, Ethiopia, India, Indonesia, Nigeria, Papua New Guinea, Peru, Poland, Thailand, and Vietnam) took part in the 12-day training in Guilin, China.

IRCK organized the training, lectures and field investigations with emphasis on the karst hydrogeology, assessment and management of karst underground water resources in China.

由联合国教科文组织国际岩溶研究中心、中国地质科学院岩溶地质研究所联合主办的"岩溶水文地质与岩溶碳循环监测"国际培训班于2010年11月29日–12月10日在桂林举办。来自巴布亚新几内亚、波兰、巴西、秘鲁、越南、印度尼西亚、印度、泰国、尼日利亚、埃塞俄比亚、中国等11个国家的17名学员,参加了为期两周的培训。

此次培训班围绕中国岩溶水文地质概况、中国岩溶地下水资源评价与管理及岩溶碳循环监测评价技术方法三个主题进行了专业的培训、讲座和野外调研活动。

The course included 28 hours of classroom instruction taught by 12 Chinese experts on karst geology and 6 experts from South Africa, Switzerland, Germany, Brazil, Slovenia and the United States.

1. Prof. Wolfgang Dreybrodt lectured on "Dissolution and precipitation kinetics of limestone".
2. Prof. Zheng Hongbo lectured on "Silicate rock weathering and the carbon cycle".
3. Dr. Thierry Bussard lectured on "Groundwater protection – general approach and vulnerability assessment in karst areas".
4. Dr. Mitja Prelovsek lectured on "Hydrogeological characteristics of classical and Dinaric karst".
5. Prof. Li Guomin lectured on "Numerical modeling for groundwater flow in karst aquifer".
6. Yuan Daoxian lectured on "Karst hydrogeology".
7. Mr. Rick Fowler lectured on "DNA analysis of bacteroides to quantify fecal contamination and identify its source".
8. Dr. Yao Yupeng lectured on "An introduction of the earth science programs in NSFC and international collaborations".
9. Prof. Augusto Sarreiro Auler lectured on "An overview of caves and karst in Brazil with emphasis in 'karst' in low solubility rocks".

18位国内外岩溶地质领域的著名专家学者应邀加入培训班讲师团，其中国内专家12名，外籍专家6名，分别来自南非、瑞士、美国、巴西、斯洛文尼亚、德国，授课时间28小时。

1. Wolfgang Dreybrodt：碳酸盐岩的溶解与沉积。
2. 郑洪波：硅酸盐的风化作用与碳循环。
3. Thierry Bussard：岩溶地区地下水保护的总体思路和脆弱性评估。
4. Mitja Prelovsek：传统岩溶和第纳尔岩溶的水文地质特征。
5. 李国敏：岩溶水文模型。
6. 袁道先：岩溶水文地质学概述。
7. Rick Fowler：类杆菌族的定量化污染DNA分析并识别其来源。
8. 姚玉鹏：国家自然科学基金地球科学计划项目的介绍和国际合作。
9. Augusto Sarreiro Auler：巴西洞穴和岩溶中尤其是低可溶性岩石溶蚀过程。

Prof. Jiang Zhongcheng explained the karst landforms in Guilin.

蒋忠诚研究员带队考察桂林岩溶地貌景观。

Dr. Wang Jinliang demonstrated the automatic karst groundwater monitoring system.

汪进良博士演示岩溶地下水自动化监测系统。

Would you please let me try the portable equipment?

便携式仪器设备在野外工作很方便,让我也试试。

Field activities included excursions to the fenglin and fengcong karst on Yao Mountain Guilin, monitoring karst carbon sink process in Maocun, and visiting the IRCK/IKG labs and the Karst Geology Museum of China. The trainees engaged the researchers and managers in lively communication about cooperation and discussion karst environmental problems.

培训班的野外实习、考察，安排了桂林尧山峰丛洼地、峰林平原景观考察、桂林毛村岩溶碳汇过程对比监测实践，同时参观了国际岩溶研究中心/岩溶所的实验室和中国岩溶地质馆等。此次培训班的学员非常活跃，利用可能的时间，与国际岩溶研究中心的研究人员和管理人员沟通、交流，寻求合作，探讨感兴趣的岩溶环境问题。

Group picture in Maocun Village.
桂林毛村实践完成后，临时"大家庭"来一个"全家福"。

During the training course, intensive communications resulted in strengthened friendships and built different degrees of cooperation among the participants. A draft MOU was drawn up between IRCK and the Vietnam Institute of Geosciences and Mineral Resources. Moreover, the Faculty of Geography, Gadjah Mada University (FGE–UGM) and IRCK signed a Memorandum of Understanding Academic Cooperation and Exchange Program towards a Partnership on Tropical Karst, education/training, and management.

1. Prof. Yuan Daoxian spoke with Mr. Chaiporn Siripornpibul from Thailand. 2. The Director of the Instituto do Carste in Brazil, Dr. Augusto Sarreiro Auler, communicating with Prof. Suyash Kumar from India. 3, 6, 8: The director of IRCK communicating with trainees from Thailand, Vietnam and Indonesia. 4. Dr. Thierry Bussard from Switzerland communicated with Prof. Wang Weiping and Song Chao. 5. The trainees learned together. 7. Dr. Eko Haryono from Indonesia and Mr. Chaiporn Siripornpibul from Thailand communicated about paleoclimate recorded by stalagmite with Prof. Zhang Meiliang from IRCK.

  本次培训在学员之间、教员与学员之间进行了广泛的交流。课间、课后开展了大量的沟通与洽谈，增进了友谊，建立了不同程度的合作意向，草拟了国际岩溶研究中心与越南地球科学与矿产资源研究所的合作意向备忘录，签署了国际岩溶研究中心与印度尼西亚噶迦玛达大学地理系的《热带岩溶研究与管理伙伴关系学术合作交流意向备忘录》。

  1：袁道先与泰国学员 Chaiporn Siripornpibul 交流；2：巴西岩溶所所长 Augusto Sarreiro Auler 与印度学员 Suyash Kumar 交流；3、6、8：国际岩溶研究中心主任分别于泰国、越南、印尼学员代表交流、洽谈；4：瑞士教员 Thierry Bussard 博士与中国学员王维平、宋超交流；5：学员间的相互学习、相互促进；7：印尼学员 Eko Haryono 与泰国学员 Chaiporn Siripornpibul 与国际岩溶研究中心张美良教授交流石笋研究方法和进展。

Group photo.

培训班结业合影。

On December 10, 2010, the closing ceremony of International Training Course on Karst Hydrogeology and Karst Carbon Cycle Monitoring was held in the IKG auditorium.

2010年12月10日，由联合国教科文组织国际岩溶研究中心、中国地质科学院岩溶地质研究所联合主办的"岩溶水文地质与岩溶碳循环监测"国际培训班结业。结业典礼在岩溶所大礼堂举行。

第七章 国际培训

Ma Yongzheng, director of the Department of International Cooperation & Science and Technology, MLR, spoke at the closing ceremony.

国土资源部科技与国际合作司马永正处长在培训班结业典礼致辞。

Dr. Suyash Kumar, head of the Department of Geology, Govt. PG Science College, Gwalior, India, gave a speech on behalf of the trainees.

印度瓜廖尔政府模式科学学院地质系主任 Suyash Kumar 博士代表学员致辞。

Prof. Augusto Auler, director of the Instituto do Carste, Brazil, gave a speech on behalf of the instructors.

巴西岩溶所所长 Augusto Auler 教授代表教员致辞。

International Training Course on Karst Hydrogeological Investigation Technology and Methodology, 2011
2011年"岩溶水文地质调查技术方法"国际培训班

The International Training Course on Karst Hydrogeological Investigation Technology and Methodology was hosted by IRCK/IKG from November 21 to December 2, 2011 in Guilin. In total 17 trainees from 9 developing countries (Brazil, Ethiopia, India, Indonesia, Nigeria, Romania, Thailand, Vietnam and China), participated in the training course. It was taught by 12 karst scientists from China and five experts from abroad (Canada, Indonesia, Switzerland, United States and UNESCO Beijing Office).

IRCK provided this two-week professional training course aimed to combine basic research with practical application, while integrating systematic lectures and reports on new frontiers in karst research. The course focused on the karst dynamic system, karst groundwater monitoring, geophysical methods and tracing techniques, hydrogeological mapping, underground water resources exploitation and management, karst processes and the carbon cycle.

由联合国教科文组织国际岩溶研究中心、中国地质科学院岩溶地质研究所联合主办"岩溶水文地质调查技术方法"国际培训班，于2011年11月21日–12月2日在桂林举办。来自巴西、埃塞俄比亚、印度、印尼、尼日利亚、罗马尼亚、泰国、越南和中国共9个国家的17位学员进行了为期两周的专业培训。

16位国内外岩溶地质领域的著名专家学者应邀加入培训班讲师团，其中国内专家12名，外籍专家5名，分别来自加拿大、印尼、瑞士、美国和联合国驻北京办事处。

此次培训以基础性和实用性并重、系统授课与前沿研究讲座相结合为原则，主要从4个方面：①岩溶水文地质与环境地质填图；②岩溶地球化学测试与分析技术和数据解译；③岩溶地下水监测与示踪技术；④地球物理方法与应用，进行专业培训、系列讲座和野外实践活动。

第七章　国际培训

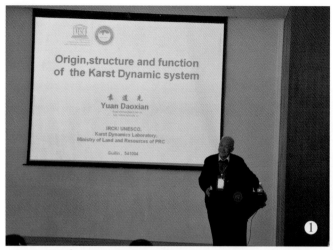

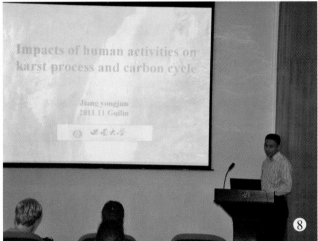

17 experts from Canada, China, Indonesia, Switzerland, the United States and the UNESCO Office in Beijing participated in this training course as lecturers. The 12 domestic and 5 foreign experts provided a total of 26 hours classroom instruction.

1. Prof. Yuan Daoxian lectured on "Origin, structure and function of the karst dynamic system".

2. Prof. Derek Ford lectured on "The empirical linear relationships between electrical conductivity and TDS in bicarbonate and sulphate waters".

3. Prof. Liu Zaihua lectured on "Geochemical variations in the karst systems of SW China: implications for the carbon cycle and environmental change study".

4. Dr. Ramasamy Jayakumar introduced the UNESCO Beijing Office and the support to IRCK.

5. Prof. Jiang Yongjun lectured on "Impacts of human activities on karst process and carbon cycle".

6. Prof. Jiang Zhongcheng lectured on "Karst water survey and exploitation in southwest China".

7. Dr. Eko Haryono lectured on "Karst management policy in Indonesia".

8. Prof. Tang Jiansheng lectured on "Large-scale karst hydrogeological mapping in field work".

9. Prof. He Shiyi lectured on "Introduction to observation and processing of hydrogeochemical data".

17位国内外岩溶地质领域的著名专家学者应邀加入培训班讲师团，其中国内专家12名，外籍专家5名，分别来自加拿大、美国、中国、瑞士、印尼、UNESCO驻北京办公室，授课时间26小时。

1. 袁道先：岩溶动力学理论与实践。
2. Derek Ford：重碳酸盐水及碳酸盐水中的电导率和溶解性固体总量之间的经验线堆关系。
3. 刘再华：中国西南岩溶系统地球化学及其与碳循环。
4. Ramasamy Jayakumar：介绍UNESCO北京办公室及相关工作和对国际岩溶研究中心的支持。
5. 蒋勇军：人类活动对岩溶过程碳汇的影响。
6. 蒋忠诚：岩溶水资源调查与开发利用。
7. Eko Haryono：印度尼西亚岩溶管理政策。
8. 唐建生：大比例尺岩溶水文地质填图野外工作方法。
9. 何师意：水文地质化学监测与示踪技术。

Field practice included tracer tests at the Maocun Karst Experimental Site, karst spring water chemistry test at the Yaji Karst Experimental Site, hydrogeological mapping and geophysical methods at the Zhaidi Hydrogeological monitoring site. The participants also visited the Karst Geology Museum of China, stalagmite room and the IRCK laboratories.

野外实践主要指在毛村岩溶实验场实际操作示踪实验、丫吉岩溶实验场水化学测试、寨底岩溶水文监测站水文地质填图方法和地球物理方法介绍，同时室内也安排了中国岩溶地质馆、石笋库和实验室参观。

右上图
Top right
Dr. Jiang Guanghui taught the students about conducting water chemistry tests at the Yaji Karst Experimental Site.
姜光辉博士带领，开展丫吉岩溶实验场水化学测试实践。

右中图
Middle right
Dr. Jiang Guanghui introduced the students to the Karst Geology Museum of China.
姜光辉讲解中国岩溶地质馆。

右下图
Bottom right
Dr. Yi Lianxing and Dr. Gan Fuping taught hydrological mapping and gave an introduction geophysical methods at the Zhaidi Hydrogeological Monitoring Site.
易连兴、甘伏平博士带领，介绍寨底岩溶水文监测站水文地质填图方法和地球物理方法。

左上图
Top left
Prof. Zhang Meiliang introduced the stalagmite collection, and spoke about the use of speleothems in climate change research.
张美良介绍石笋库及最新利用石笋开展过去气候变化研究进展。

左中图
Middle left
Dr. Wang Jinliang taught tracer test methods at the Maocun Karst Experimental Site.
汪进良博士带领，介绍毛村岩溶实验场示踪实验实际操作方法。

左下图
Bottom left
Dr. Zhang Chunlai introduced the atomic absorption spectrometer.
张春来博士介绍原子吸收光谱仪。

右上图
Top right
Nguyen Xuan Nam from Vietnam gave a report on "Approach to karst area studying by cave expedition".
越南学员 Nguyen Xuan Nam 做"利用探洞研究岩溶地区的方法"报告。

右中图
Middle left
Luciana Alt from Brazil gave a report on "Overview on effectiveness of the Lagoa Santa Karst Protected Area".
巴西学员 Luciana Alt 做"巴西拉哥亚圣塔岩溶保护地区有效性综述"报告。

右下图
Bottom left
Demis Alamirew Ayenew from Ethiopia gave a report on "Hydrogeological Overview some of Parts of Ethiopia".
埃塞俄比亚学员 Demis Alamirew Ayenew 做"埃塞俄比亚部分地区的水文地质综述"报告。

第七章 国际培训

During the training course, each student gave a presentation as part of their training assessment. The presentation gave the students an opportunity to share their research and ideas, and engage in group learning.

通过此次培训班,学员们对评估工作做了充分的准备,达到了相互交流、教学互长的效果。

左上图
Top left

Emilya Nurjani from Indonesia gave a report on "Characteristcs of spring in Wonogiri karst region, Central Java".

印度尼西亚学员 Emilya Nurjani 做"爪哇中部沃诺吉里岩溶地区泉水的特征"报告。

左中图
Middle left

Shitta Kazeem Akorede from Nigeria gave a report on "Overview on Lithostratigraphy of Nigeria".

尼日利亚学员 Shitta Kazeem Akorede 做"尼日利亚地区岩相层序综述"报告。

左下图
Bottom left

Samad John Smaranda from Romania gave a report on "Karst ecosystems protection in Romania".

罗马尼亚学员 Samad John Smaranda 做"罗马尼亚岩溶生态系统的保护"报告。

Training course graduation group photo.
培训班结业合影。

The closing ceremony of the training course was held in the IKG auditorium on December 2. Dr. Ramasamy Jayakumar from the UNESCO Beijing Office; Prof. Lian Changyun, Deputy Director of the Department of Foreign Affairs for Science and Technology, CGS; and CAGS Executive Vice President Zhu Lixin took part in the ceremony and gave speeches. Prof. Derek Ford spoke on behalf of the instructors. Prof. Yuan Daoxian, representing IRCK, gave a summary report for the training course.

12月2日下午在岩溶所大礼堂，举行培训班结业仪式。特邀嘉宾Jayakumar、中国地质调查局科外部连长云主任、中国地质科学院常务副院长朱立新等参加闭幕式，并致辞；Derek Ford教授代表教员致辞；袁道先院士代表国际岩溶研究中心做培训班总结发言。

The UNESCO Beijing Office Mr. Jayakumar took part in the ceremony and gave a speech.
联合国教科文组织北京办事处 Jayakumar 参加闭幕式，并致辞。

Prof. Lian Changyun, Deputy Director of the Department of Foreign Affairs for Science and Technology, CGS took part in the ceremony and gave a speech.
中国地质调查局科外部主任连长云参加闭幕式，并致辞。

CAGS Executive Vice President Zhu Lixin took part in the ceremony and gave a speech.
中国地质科学院常务副院长朱立新教授参加闭幕式，并致辞。

Derek Ford gave a speech on behalf of the instructors.
Derek Ford 教授代表教员致辞。

International Training Course on Karst and Hydrogeochemistry, 2012
2012 年"岩溶与水文地球化学"国际培训班

The 4th IRCK International Training Course on Karst and Hydrogeochemistry was co-sponsored by the Southwest University of China, the UNESCO Beijing Office and IGCP/SIDA 598. It was held in Chongqing, China from November 25 to December 7, 2012. The 20 trainees from 11 countries (Hungary, Slovenia, Romania, Vietnam, Brazil, Ethiopia, Malaysia, Thailand, India, Indonesia and Kenya) participated in the course. Lectures were given by 21 karst scientists from Brazil, China, Slovenia, Spain, and the United States. Topics included evaluation and modeling of karst hydrochemistry data; carbon, water and calcium monitoring methods and data interpretation for karst dynamic system; stable isotope geochemistry; land use and hydrogeochemistry.

由国际岩溶研究中心主办，西南大学、联合国教科文组织北京办事处和国际地学计划项目 IGCP/SIDA 598 协办，"岩溶与水文地球化学"国际培训班于 2012 年 11 月 25 日 –12 月 7 日在重庆举办，来自匈牙利、斯洛伐克、罗马尼亚、越南、巴西、埃塞俄比亚、马来西亚、泰国、印度、印度尼西亚、肯尼亚等 11 个国家的 20 名学员进行了为期两周的专业培训。来自斯洛文尼亚、巴西、美国、西班牙、中国等 5 个国家的 21 名教员参加了培训班。

此次培训班比较系统地讲授了岩溶水化学数据分析、评价与建模、碳水钙监测方法与数据解译、稳定同位素地球化学、土地利用与水化学等内容。

1. Yuan Daoxian–Climate Change and Karst Hydrogeology.
2. Ben Miller–Methodologies & Case Studies of Groundwater Tracing Application in Karst Areas.
3. Vitor Moura–Monitoring procedures for management and protection of caves in Brazil experiences and challenges.
4. Tadej Slabe–Karstology.
5. Bartolome Andreo–Protecting groundwater in karst media.
6. Martin Knez–Planning traffic roads crossing karst.
7. Zhang Dian–Causality analysis of Climate change and large-scale human crisis during historical time.
8. Yang Yan–Environmental Isotope Introduction: Techniques and Applications.
9. Qi Shihua–Persistent Organic Pollutants in Karst Areas.

1. 袁道先院士授课"全球变化与岩溶水文地质问题"。
2. Ben Miller 教授授课"岩溶地区地下水示踪应用技术方法与案例研究"。
3. Vitor Moura 教授授课"巴西洞穴管理与保护监测规程面临的经验与挑战"。
4. Tadej Slabe 教授授课"岩溶学发展"。
5. Bartolome Andreo 教授授课"岩溶地下水保护"。
6. Martin Knez 研究员授课"穿越岩溶地区交通道路规划"。
7. 章典教授授课"历史上气候变化与大规模人类灾难的因果分析"。
8. 杨琰教授授课"环境同位素基本原理、实验方法和实践应用"。
9. 祁士华教授授课"岩溶地区持久性有机污染物研究"。

第七章 国际培训

国际岩溶研究中心 6 年历程

Chapter 7 International Training

第七章 国际培训

The training course included visits to laboratories of the School of Geographical Sciences of Southwest University and Chongqing Geological Museum. Students also experienced World National Heritage Site, Wulong National Geopark, Jinfo Mountain in Nanchuan district, Chongqing. In Nanchuan, trainees and lecturers had a communication with local experts and administrators.

配合此次国际培训，组织方不仅组织参观了西南大学实验室、重庆市地质博物馆，还组织学员考察了世界自然遗产地、武隆国家地质公园、南川区金佛山，并与当地专家、管理人员进行了交流。

对页左上图
Oppsite page, top left
Students visited laboratories at the School of Geographical Sciences, Southwest University.
参观西南大学实验室。

对页右上图
Oppsite page, top right
Trainees communicated with local experts and administrators in the Nanchuan karst area.
与南川区专家、管理人员交流。

对页下图
Oppsite page, bottom
Visit to Jinfo Mountain in Nanchuan district, Chongqing.
考察南川金佛山。

上图
Top
Visit to Furong Cave in Wulong county, Chongqing.
参观武隆芙蓉洞。

下图
Bottom
Visit to Chongqing Geological Museum.
参观重庆市地质博物馆。

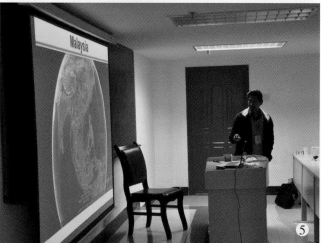

The trainee assessment gradually evolved into a platform for sharing new ideas and communicating new needs and trends in karst research.

1. Alena Petrvalská from Slovakia introduced the ecological restoration and evolution process in Slovakia karst area in recent 60 years.

2. Alexandra Hillebrand-Voiculescu from Romania introduced the relationship between microorganism in cave glaciers and climate change in the past.

3. Mahippong Worakul from Thailand introduced the karst hydrogeology in Thailand.

4. Dang Tran Nhu Thuy from Vietnam introduced karst landscape and ecological environment problems included rocky desertification problem on karst areas in the Ha Long Bay Natural Heritage and karst area in Jianghe province, Vietnam.

5. Nurzaidi Abdullah from Malaysia introduced the karst in Malaysia.

6. Gabriella Koltai from Hungary introduced the Hungarian karst areas.

7. Marina Leão from Brazil introduced the iron caves in Brazil.

8. Luiz Eduardo Panisset Travassos from Brazil introduced the possibilities of karst and cave research in Minas Gerais, Brazil.

9. Abrham Gebreslassie Gebre from Ethiopia introduced water purification technology in Ethiopia and serious karst water pollution problems from Improper waste disposal.

学员评估逐渐演化为启迪岩溶研究新思路、交流岩溶研究新需求、新趋势的平台。

1. 斯洛伐克学员介绍了斯洛伐克岩溶区近60年以来生态恢复、演变过程。

2. 罗马尼亚学员Alexandra Hillebrand介绍了洞穴冰川中微生物与过去气候变化的关系。

3. 泰国学员Mahippong Worakul介绍了泰国的岩溶水文地质概况。

4. 越南学员Dang Tran Nhu Thuy介绍了越南下龙湾自然遗产与河江省岩溶区的岩溶景观、生态环境问题，包括石漠化问题。

5. 马来西亚学员Nurzaidi Abdullah介绍了马来亚的岩溶现象。

6. 匈牙利学员Gabriella Koltai介绍了匈牙利的岩溶区域。

7. 巴西学员Marina Leão介绍巴西铁洞。

8. 巴西学员Luiz Eduardo Panisset Travassos介绍了巴西米纳斯基拉斯地区岩溶和洞穴研究的可能性。

9. 埃塞俄比亚学员Abrham Gebreslassie Gebre则分别分析了本国水净化技术和不当废弃物处理带来的严重岩溶水污染问题。

On December 6, 2012, the closing ceremony of the training course was held at Southwest University. CAGS executive deputy President Wang Xiaolie fully affirmed that IRCK had realized its target of disseminating the newest ideas and methods in karst hydrogeology for the betterment of water resources and for addressing environmental problems in karst areas. Prof. Wang stated that CAGS will continue to support the coordination work between IRCK and UNESCO, and the development of IRCK.

Photo: Prof Yuan Daoxian, Prof. Wang Xiaolie, Jiang Shijin, Dr. Ramasay Jayakumar, Cui Yanqiang and Prof. Jiang Yuchi awarded certificates to the trainees.

2012年12月6日，培训班结业典礼在西南大学举行，中国地质科学院党委书记王小烈在致辞中，充分肯定了国际培训班对中心国际化目标的实现、参与国际岩溶水文地质最新知识的传播，解决岩溶区普遍面临的水资源与环境问题能力的提高，加深相互了解和开展进一步合作均具有积极作用，并表示，地科院今后将继续做好国际岩溶研究中心与 UNESCO 的协调工作，关心和支持中心建设与发展。

最后，嘉宾袁道先、王小烈、蒋仕金、贾古玛 (Jayakumar)、崔延强和姜玉池为学员颁发了结业证书。

对页图 / Oppsite page
Distinguished guests awarded certificates to the trainees.
嘉宾为学员颁发结业证书。

上图 / Top
From left to right: Dr. Ramasay Jayakumar, Prof. Wang Xiaolie, Cui Yanqiang, and Prof. Jiang Yuchi gave speeches at the closing ceremony.
从左到右：贾古玛、王小烈、崔延强、姜玉池在结业典礼上致辞。

International Training Course on Karst Hydro-geological Survey, Dynamic Monitoring and Application in River Basins, 2013
2013年"流域岩溶水文地质调查、动态监测与应用"国际培训班

The Fifth International Training Course on Karst Hydro-geological Survey, Dynamic Monitoring and Application in River Basins was organized by IRCK from November 17 to 29, 2013 in Guilin, China.

21 trainees from 18 countries (China, Brazil, Nigeria, Mexico, Vietnam, Malaysia, Uganda, Thailand, Zimbabwe, Iran, Slovenia, Indonesia, Romania, Hungary, Slovakia, Russia, US, and South Africa) took part in the 12-day training course. 16 experts from six countries (Iran, US, China, South Africa, Germany and Serbia) joined the course as facilitators.

The training course focused on five topics:
1) karst hydrological survey and monitoring approaches;
2) geophysical method of defining karst river basin boundaries;
3) design and construction of monitoring sites;
4) river basin water resources assessment; and
5) karst processes and carbon sink effects.

In June 2013, the international training course was approved as an IKG project and the project's implementation plan was completed. In September, documents were submitted to CAGS for project approval, and the recruitment/enrollment of trainees began. In October, the course agenda and participant list were initially determined, the relevant procedures for foreign participants to visit China were implemented, and the foreign trainers' visits to China were scheduled. In November, a preparation meeting for the course was held, the work management system and the roles and responsibilities for the work group and staff were defined, and training materials were produced.

There were five components of the course:
1) classroom lectures (six days);
2) visits (Karst Geology Museum of China, Stalagmite Archive, and isotope, water chemistry and mass spectroscopy laboratories, half a day);
3) field practice (at Yaji Karst Experimental Site, and Zhaidi Karst Hydrological Monitoring Station, two days);
4) trainees' assessment (one and a half days); and
5) sight-seeing on weekends (recommended excursion along Lijiang River and city tour in Guilin, two days).

Field practice included a tracer experiment at Maocun Karst Experiment Site, water chemistry tests at Yaji Karst Experimental Site, and hydrogeological mapping and geophysical observation methods at Zhaidi Karst Hydrological Monitoring Station.

11 Chinese and 5 foreign researchers served as course trainers/facilitators.

Chinese trainers/facilitators: Yuan Daoxian, Jiang Zhongcheng, Chen Weihai, Gan Fuping, Yi Lianxing, Zhang Cheng,

He Shiyi, Li Qiang, Jiang Guanghui, Wei Yuelong, and Zhao Liangjie;

　　Foreign trainers/facilitators: Ezzat Raeisi (Iran), George Veni (US), Harrison Hursiney Pienaar (South Africa), Jonathan D. Arthur (US) and Zoran Stevanovic (Serbia).

　　2013年11月17-29日国际岩溶研究中心于桂林举办"流域岩溶水文地质调查、动态监测与应用"第5次国际培训班。来自巴西、尼日利亚、墨西哥、越南、马来西亚、乌干达、泰国、津巴布韦、伊朗、斯洛文尼亚、印尼、罗马尼亚、匈牙利、斯洛伐克、俄罗斯、美国、南非等18个国家的21名学员参加为期12天的培训。来自伊朗、美国、中国、南非、德国、塞尔维亚等6个国家的16名专家，作为教员参加了培训班。

　　此次培训的主要内容为岩溶水文调查、监测方法，岩溶流域边界的确定和地球物理方法，监测站的设计与建设，流域水资源评价，岩溶作用与碳汇效应。

　　国际培训班项目在2013年6月获批中国地质科学院岩溶地质研究所所控项目；7月完成培训班项目实施方案；9月上报主管部门（中国地质科学院）项目审批材料，递交请示文件，下旬开展招生工作；10月基本确定培训日程及培训参与人员名单，办理外籍人员来华相关手续，确定教员来华时间；11月召开培训班专题筹备会，明确工作管理制度、工作小组、工作人员及相关职责，确定培训教材。

　　此次培训主要包括5个部分：课堂讲学（6天）；参观（岩溶博物馆、石笋库、同位素、水化学、质谱测定实验室，共0.5天）；野外实践（丫吉岩溶实验场，寨底岩溶水文监测站，2天）；学员评估（1.5天）；周末休息（推荐漓江游览和桂林市内观光，2天）。

　　野外实践主要指在毛村岩溶实验场实际操作示踪实验、在丫吉岩溶实验场开展水化学测试，在寨底岩溶水文监测站介绍水文地质填图方法和地球物理方法。

　　参加本次培训班的讲师包括国内讲师11名、外籍讲师5名。

1. Academician Yuan Daoxian lecturing on "Origin, structure and function of karst dynamic system".

2. Prof. Zhang Cheng lecturing on "Carbon cycle of karst system and its potential contribution to atmospheric carbon sink".

3. Prof. Jiang Zhongcheng lecturing on "Water conservancy projects construction in karst areas".

4. Prof. Ralf Kaldenhoff lecturing on "Removing atmospheric CO_2 concentration with algae techniques".

5. Prof. Jonathan D. Arthur lecturing on "Types of sinkholes in Florida".

6. Dr. Harrison Hursiney Pienaar lecturing on "Sustainable water supply on semi-arid areas: resource protection measures and management interventions".

7. Dr. George Veni lecturing on "Environmental impact assessment".

8. Prof. Zoran Stevanovic lecturing on "Methods of water balance and storage assessment of karst aquifers".

前页图 / Front page

Participants attending the Fifth IRCK International Training Course in 2013.
2013 年国际岩溶培训班合影。

1. 袁道先院士在做"岩溶动力系统的起源、结构和功能"的报告。
2. 章程研究员在做"岩溶系统的碳循环和对大气碳汇的潜在贡献"的报告。
3. 蒋忠诚研究员在做"岩溶地区的水利建设"的报告。
4. Ralf Kaldenhoff 教授在做"藻类技术对大气二氧化碳浓度的减少"的报告。
5. Jonathan D. Arthur 教授在做"佛罗里达州的塌陷类型"的报告。
6. Harrison Hursiney Pienaar 博士在做"半干旱区可持续的供水系统：资源保护对策和管理法"的报告。
7. George Veni 教授在做"环境影响评价"的报告。
8. Zoran 教授在做"岩溶含水层水分平衡和储量评估方法"的报告。

Chapter 7 International Training

Visiting Zhaidi Karst Hydrological Monitoring Station
考察寨底岩溶水文地质基地

Visiting Yaji Karst Experimental Site
考察丫吉岩溶实验场

Participants attending the Fifth IRCK International Training Course visiting the Zhaidi Karst Hydrological Monitoring Station
2013年国际岩溶培训班在寨底岩溶水文地质基地合影

During the international training course, IRCK conducted friendly exchanges and negotiations with research institutions from many countries, seeking practical cooperative means and projects, including:

1. FGS Director Jonathan Arthur met with IRCK Director Jiang Yuchi

Dr. Jonathan Arthur, director of the Florida Department of Environmental Protection's Florida Geological Survey, US, was invited to serve as a trainer/facilitator for IRCK's 2013 course. During his stay he met with IRCK Director Jiang Yuchi to discuss future cooperation and related issues.

Both parties exchanged ideas on cooperation in sinkhole mapping and carbon sink monitoring and research. Dr. Arthur expressed his strong support for an IRCK karst carbon sink monitoring and research site in Florida, and offered to provide background information and possibly human resources. The details of the cooperation have yet to be defined.

上图
Top

Dr. Jonathan Arthur meeting with IRCK leaders.
Jonathan Arthur 与 IRCK 交流。

在国际岩溶研究中心国际培训班期间，与多国研究机构进行了友好交流和协商，探寻可操作的合作内容和合作方式。主要包括：

1. 国际岩溶研究中心主任姜玉池先生与美国佛罗里达州地质调查局局长 Jonathon Arthur 先生会谈

美国佛罗里达州地质调查局局长、佛罗里达州首席地质学家乔纳森·亚瑟（Jonathan Arthur）先生应邀作为第 5 届国际岩溶研究中心培训班教员，期间，亚瑟先生与姜玉池主任会面讨论了未来开展项目合作的相关事宜。

通过交流在岩溶塌陷调查填图、碳汇监测研究等领域有合作的前景。美方表示欢迎和积极支持国际岩溶研究中心在佛罗里达州建设岩溶碳汇监测研究站点，并可提供背景资料和可能的人力资源，具体的合作细节、操作过程需进一步落实。

2. University of Belgrade, Serbia delegation visited IRCK

In late November 2013, Prof. Zoran Stevanovic, director of the Centre for Karst Hydrogeology at the University of Belgrade, Serbia visited IRCK/IKG. During his visit he served as a member of the expert panel appointed by UNESCO to conduct the First Six-Year Assessment of IRCK, and as a trainer/facilitator of the Fifth IRCK International Training Course.

Prof. Stevanovic spoke with Prof. Jiang about issues related to implementing the cooperative agreement signed by both parties in 2011, particularly the karst groundwater monitoring and mathematical modeling. Prof. Stevanovic expressed strong interest in building karst carbon sink monitoring stations in Serbia, and invited Prof. Jiang to attend the international conference and field seminar "Karst Without Boundaries" to be held in Bosnia in 2014 and to visit the Centre for Karst Hydrogeology during the summer meeting of DIKTAS. Prof. Jiang accepted the invitation with pleasure and thanked Prof. Stevanovic for sending IRCK several academic monographs, particularly Climate Changes and Impacts on Water Supply and Water Resources & Environmental Problems in Karst.

上图
Top

Dr. George Veni, director of NCKRI and Prof. Jiang Yuchi, director of IRCK/IKG, signing a cooperative agreement.

美国国立洞穴与岩溶研究所所长乔治·维尼（George Veni）先生与国际岩溶研究中心姜玉池先生签署合作协议。

2. 与塞尔维亚贝尔格莱德大学 Zoran Stevanovic 教授交流

塞尔维亚贝尔格莱德大学岩溶水文地质中心主任 Zoran Stevanovic 教授应邀参加 UNESCO 为 IRCK 第一个六年评估的专家评估组成员及第 5 届国际岩溶研究中心培训班教员，于 11 月下旬访问国际岩溶研究中心 / 岩溶所。

访问期间，与国际岩溶研究中心主任、岩溶所所长姜玉池教授就 2011 年签署的合作协议条款的落实进行了磋商，尤其在岩溶地下水监测、数学模型构建方面。Zoran Stevanovic 教授非常愿意在塞尔维亚构建岩溶碳汇监测站，并热情邀请姜玉池主任明年在参加波黑"Karst without Bouadary"国际研讨会及 DIKTAS 总结大会期间，访问贝尔格莱德大学岩溶水文地质中心。姜玉池主任愉快地接受了邀请，同时对 Zoran Stevanovic 教授赠送给国际岩溶研究中心的学术专著非常感谢，尤其是气候变化及其对水资源供应的影响，岩溶地区水资源及环境问题。

3. NCKRI Executive Director George Veni signed an agreement with IRCK/IKG

Dr. George Veni, executive director of the US National Cave and Karst Research Institute (NCKRI), visited IRCK/IKG in late November, 2013. He served as a member of the expert panel appointed by UNESCO to conduct the First Six-Year Assessment of IRCK and as a course trainer/facilitator. During his visit, Dr. Veni spoke with Prof. Jiang and signed a cooperative agreement with IRCK/IKG.

Dr. Veni gave an overview of NCKRI. He mentioned a US project to digitize and archive many scientific monographs or journals on karstology, including many from his private library. He proposed adding Carsologica Sinica sponsored by IKG, to this online reference library on karst science. He also hoped IKG can co-organize the Multidisciplinary Conference on Sinkholes and the Engineering and Environmental Impacts of Karst in 2017.

Prof. Jiang thanked Dr. Veni for participating in IRCK activities, and for providing support and input for IRCK's development. He was also glad to learn that Dr. Veni has been elected as a member of the Second IRCK Academic Committee. He noted that both the US and China are rich in karst resources, and the karst scientists from the two countries have been working closely for a long time, believing the cooperation between the two parties will embrace a brighter future.

Both parties signed a cooperative agreement between NCKRI and IRCK/IKG.

3. 美国国立洞穴与岩溶研究所所长乔治·维尼与我所签订合作协议

美国国立洞穴与岩溶研究所所长乔治·维尼（George Veni）先生应邀作为国际岩溶研究中心第一个六年评估专家组成员，第五届国际岩溶研究中心培训班教员，于 11 月下旬访问国际岩溶研究中心 / 岩溶地质研究所。期间，维尼先生与姜玉池主任进行了合作交流，并签署了合作协议。

维尼先生介绍了美国国立洞穴与岩溶研究所的基本情况，特别提到维尼先生的私人图书馆里珍藏了包括我所主办的《中国岩溶》在内的多种岩溶学专著或期刊，期望逐步构建为比较齐全的岩溶科学文献资料库。此外，维尼先生表示，他希望我所能协办 2017 年的岩溶塌陷和岩溶工程与环境影响多学科国际研讨会。

4. Dr. Bogdanov from Russia conducted talks with IKG on future cooperation in karst science

Representatives from IKG and Russia's Geologic Research Institute for Construction held talks on future cooperation and exchange in karst science on November 27, 2013. The meeting was attended by Prof. Jiang, researchers from relevant departments within IKG, and Dr. Mikihail Bogdanov, director of Russia's Geologic Research Institute for Construction. This marked ongoing development of international collaboration and exchange in karst science under the framework of IRCK.

Prof. Jiang opened the meeting by extending a warm welcome to Dr. Bogdanov on behalf of IKG, and introduced IKG's major research topics and outstanding achievements in scientific research and geological survey in recent years. Then Dr. Bogdanov spoke highly of IKG's important role in providing

Dr. Mikihail Bogdanov meeting with IRCK leaders.
Mikihail Bogdanov 博士与 IRCK 成员合影留念。

support to the operation of IRCK, and presented the information about the types and distribution of karst in Russia, as well as the technical progress of the Geologic Research Institute for Construction in monitoring karst geological disasters. Both parties agreed that implementing international cooperation and exchange in karst science based on IRCK is of vital importance to promote karst research by both parties. It was also agreed that both parties will enhance communication, understanding and cooperation. In 2014, both parties will send delegations to each other's organization to conduct study tours and exchange activities on karst geology and related fields, and sign a framework agreement on research project collaboration on the basis of full understanding and mutual trust.

Finally, Prof. Jiang thanked Dr. Bogdanov for attending the Fifth IRCK International Training Course, and invited him to participate in the 2014 course. Dr. Bogdanov accepted the invitation with pleasure and thanked Prof. Jiang and IKG for hosting the event and their kind hospitality.

4. 与俄罗斯联邦工程建设地质研究所 Mikihail Bogdanov 博士交流

2013 年 11 月 27 日上午，中国地质科学院岩溶地质研究所与俄罗斯联邦工程建设地质研究所在广西桂林就未来在岩溶领域开展合作交流举行会谈，岩溶地质研究所所长姜玉池研究员、俄罗斯工程建设地质研究所所长 Mikihail Bogdanov 博士以及有关部门负责同志参加了本次会谈。

姜玉池研究员代表岩溶地质研究所，对 Mikihail Bogdanov 所长的到访表示热烈的欢迎，并介绍了岩溶地质研究所的主要研究方向和近年来在科研地调方面取得的成果。Mikihail Bogdanov 所长高度评价了岩溶地质研究所为国际岩溶研究中心运行提供的重要支撑作用，并介绍了俄罗斯的岩溶类型和分布情况，以及俄罗斯联邦工程建设地质研究所在岩溶地质灾害监测方面的技术进展。双方一致认为，依托国际岩溶研究中心，开展岩溶领域的国际合作与交流，对提升双方的岩溶研究水平具有重要的意义。双方一致同意加强沟通了解和业务往来，2014 年，双方将互派代表团开展岩溶地质和相关领域的考察和交流，并在充分了解互信的基础上，签署双方项目合作研究框架协议。

姜玉池所长对 Mikihail Bogdanov 所长来参加国际岩溶研究中心培训班表示衷心感谢，并邀请俄罗斯联邦工程建设地质研究所同事参加下年度的国际岩溶研究中心国际培训班，Mikihail Bogdanov 所长愉快地接受邀请。

Photo caption:
Name: Guangxi Luzhai Xiang Bridge National Karst Geopark
Location: DazhaoVillage,Zhongdu Town, Luzhai county, Guangxi
Time as China National Geopark: 2005
Summary: Xiang Bridge is nice Geopark combined with subtropical karst landscape and ecological landscape. And there are some endangered plant species and some Ficusmicrocarpa with age over thousand years, etc. The Geopark Characterized as natural bridge, valley, waterfall and wonderful cave-Nine Dragon Cave.

照片说明：
名称：广西鹿寨香桥喀斯特地质公园
所在地点：广西鹿寨县中渡镇大兆村
入选中国地质公园时间：2005年
概述：以融亚热带岩溶地貌景观和生态景观为一体的地质公园。公园内植被丰富，有重点保护植物岩溶山花和树龄逾千年的古榕等。公园中的香桥天生桥、香桥岩溶峡谷、响水低头瀑布，以及九龙洞中的脑纹状洞穴沉积等景观，是我国地貌地质景观中的精品。

Chapter 8

Summary

第八章 总结

Chapter 8 Summary

This chapter is based on the Agreement between the People's Republic of China and the United Nations Educational, Scientific and Cultural Organization (UNESCO) concerning the establishment and operation of International Research Center on Karst in Guilin, China, under the auspices of UNESCO, and Medium-term Strategy during 2008-2013, the 34C/4 protocol, to make a summary.

本章根据 2008 年 2 月，中国政府与联合国教科文组织签署的关于《中华人民共和国政府与联合国教育、科学及文化组织关于在桂林建立由教科文组织赞助的国际岩溶研究中心及其运作的协定》及教科文组织 2008 – 2013 年中期规划和对策文件 34C/4 的相关要求，对照 2008 年国际岩溶研究中心成立以来的工作做一简短的总结，有利于专家、评委对国际岩溶研究中心未来的建设和发展提出建议。

Agreement between the People's Republic of China and The United Nations Educational, Scientific and Culture Organization (UNESCO) concerning the establishment and operation of the International Research Centre on Karst in Guilin, China, under the auspices of UNESCO

The Government of the People's Republic of China AND The United Nations Educational, Scientific and Cultural Organizaiton (hereinafter called "UNESCO")

Preamble

Bearing in mind the Constitution of the United Naitons Educational, Scientific and Cultural Organization, adopted on 16 November 1945,

Whereas the UNESCO General Conference, at its thirty-fourth session (Resolution 34C/32) decided that the International Research Centre on Karst (hereinafter referred to as "the Centre") would be established in Guilin, the People's Republic of China, under the auspices of UNESCO,

Whereas the Government of the People's Republic of China has contributed and stands ready to contribute further to the establishment and operation of the Centre on its territory,

Welcoming the effective measures already taken by the Chinese Academy of Geological Sciences with a view to establishing the Centre and assuring the necessary conditions for the proper operation of the Centre within the framework of the Chinese Academy of Geological Sciences,

Considering that in response to an initial proposal by the Government of the People's Republic of China to establish on its territory an international centre to study the karst regions under the auspices of UNESCO, the 35th session of the International Geoscience Programme (IGCP) Scientific Board adopted Resolution IGCP 35/1 which welcome the establishment of the Centre,

Noting also with appreciation the readiness of the Institute of Karst Geology, Chinese Academy of Geological Sciences to contribute, both materially and otherwise, to the establishment and operation of the Centre,

Desiring to set forth the conditions and modalities of the cooperation regarding both the establishment and the activities of the Centre,

Considering that the Director-General of UNESCO has been authorized by the General Conference of UNESCO to conclude with the Government of the People's Republic of China an Agreement in conformity with the draft which was submitted to the General Conference. Desirous of defining the terms and conditions governing the contribution that should be granted to the said Centre in this Agreement.

The Main Advances of IRCK in the First Six Years
国际岩溶研究中心第一个6年的主要进展

The objectives and functions of the IRCK
国际岩溶研究中心的目标与功能

The objectives of the IRCK are to:

(1) Advance karst dynamics through scientific research, publications, and international cooperation.

(2) Advance international cooperation and contacts and provide a platform for the exchange of scientific information about karst dynamics and sustainable utilization of karst resources and eco-environmental protection between institutions worldwide within the framework of the International Geoscience Programme (IGCP) of UNESCO, the karst committee of the International Union of Geological Sciences (IUGS) and the International Association of Hydrogeologists (IAH).

(3) Provide advisory activities, technical information and training as a basis to develop and implement new integrated methods of rock desertification rehabilitation and ecological restoration; promote social awareness-raising within the scope of karst dynamics application for integrated methods of rock desertification rehabilitation including: society at large, non-Governmental Organizations (NGOs) and governmental institutions at central and regional levels.

(4) Develop a network of demonstration sites for the implementation of the karst dynamic system theory to improve epikarst and subterranean water resources utilization rate and protection, create positive socio-economic feedback and provide relevant ecosystem services.

(5) Promote advanced scientific research on karst dynamics, monitoring and modeling systems, as well as transfer of knowledge and its implementation in order for karst fragile environment to be ecologically sound, and rehabilitation and restoration of karst ecosystems, and sustainable development of karst regions.

国际岩溶研究中心的目标为：

（1）通过科学研究、出版活动和国际合作，促进岩溶动力学的发展。

（2）促进国际合作与交往，为世界各地属于教科文组织国际地球科学计划、国际地质学联盟（IUGS）岩溶委员会以及国际水文地质学家协会（IAH）岩溶委员会范围内的机构搭建一个有关岩溶动力学、岩溶资源可持续利用和生态环境保护的科学信息交流平台。

（3）提供咨询服务、技术信息和培训，为制定和实施新的石漠化治理和生态恢复综合方案奠定基础；在岩溶动力学应用的范围内，提高社会（包括社会大众、非政府组织以及中央和地区的政府机构）对石漠化综合治理方案的认识。

（4）为岩溶动力系统理论的实施建立示范基地网，以提高表层岩溶水和地下水资源的利用率，引起积极的社会经济反馈，并提供相关的生态系统服务。

（5）促进岩溶动力学、监测和建模系统、知识转让及其实施方面的先进科学研究，以保持脆弱的岩溶环境的良性生态循环，促进岩溶生态系统的治理与恢复，以及岩溶地区的可持续发展。

国际岩溶研究中心的功能为：

（1）开展现代岩溶学方面的实验和理论科学研究；

（2）建立并加强机构网和信息网，以方便在国际上交流科技信息和政策信息，协调并组织国际地区间的合作项目，提供实验设施和现场实验基地；

（3）从事国际技术咨询，举办国际专题讲习班和研讨会，支持野外科研调查，组织巡回讲学团；

（4）与国际水文计划（IHP）、人与生物圈计划（MOST）、社会科学及人文科学部门社会变革管理计划（SHS-MOST）、政府机构、非政府组织和决策者合作，以便将科研成果付诸应用；

（5）参与石漠化研究的协调与组织，进行人员培训和信息交流，提供相关项目的咨询服务，促进生态教育，提高公众对石漠化治理、生物多样性和可持续发展之间相互关系的意识。

The functions of the IRCK shall be to:

(1) conduct experimental and theoretical scientific research on modern karstology;

(2) create and reinforce institutional and information networks for the exchange of scientific, technical and policy information at the international level, coordinate and organize international/interregional cooperative projects and provide experimental facilities and field experimental bases;

(3) undertake international technical consultation, organize international workshops and symposia on special subjects, scientific field investigations and lecture tours;

(4) cooperate with the International Hydrological Programme (IHP), the Man and the Biosphere (MAB) Programme, the Intergovernmental Oceanographic Commission (IOC), the Social and Human Sciences-Management of Social Transformations Programme (SHS-MOST), government agencies, NGOs, institutions and decision-makers in order to put the results of scientific research into practice;

(5) participate in the coordination and organization of studies on rock desertification and ecological restoration, and conduct personnel training, exchange of information and consulting services for related projects, promote ecological education and increase public awareness of the links between rock desertification rehabilitation, biodiversity and sustainable development.

The IRCK makes efforts to achieve its objectives and perform functions
国际岩溶研究中心建设对目标的实现与功能发挥

Depending on the Agreement, IRCK shall pursue the above objectives and functions in close cooperation with IGCP and other water-related and ecosystem-related centres under the auspices of UNESCO.

With the objectives and functions of IRCK in mind, we can summarize the work of IRCK in first 6 years, as the below 4 points.

国际岩溶研究中心必须与国际地球科学计划和由教科文组织赞助的其他与水和生态系统相关的中心密切配合，努力实现上述目标，履行上述功能。

对照中心的目标、功能，将国际岩溶研究中心6年来的工作总结如下4点。

Karst Dynamic System Research and Extension
岩溶动力系统研究与推广

IRCK has been concerned about karst dynamic system research since it established, and globally to actively promote new ideas for modern karst. It is mainly summarized as the following 4 aspects:

① **Scientific research:** Following the UNESCO strategy for action on climate change International research center on karst is particularly concerned about the research fields: the karst dynamic system and the carbon cycle, stalagmite recording on paleocliamte reconstruction and karst hydrogeology, water environment and water resources exploitation and utilization.

We have successfully launched 26 finished and ongoing projects, 8, 3, 8, 2 and 5, related to kartstification and carbon cycle, stalagmite recording on paleocliamte reconstruction, karst water, calcium cycle in karst and sustainable development, respectively. IRCK had also successfully organized relevant experts to go to the severe drought area in southwest China in 2010, Shandong in 2011, and Yunnan in 2012. Large-scale pre-karst hydrogeology survey data was obtained and combined with geophysical exploration, completed the tasks of finding ground water resources.

② **Publicity and promotion:** The definition of karst dynamic system is printed on the IRCK annual report every year. Prof. Yuan Daoxian, director of the IRCK academic committee, always do the introductory report "Origin, structure and function of the Karst Dynamic system" in the IRCK training courses every year.

③ **Publication:** In addition to actively publish research papers in academic journals, IRCK also specifically coauthored with Chinese Science Bulletin, to make Special Issue entitled "Geological Processes in Carbon Cycle" (Volume 56, Number 36, December, 2011). Moreover, IRCK also coauthored with "Man and Biosphere", Special Issue entitled "Rocky Desertification in Karst Area" (Series No.59, May, 2009). The monograph was entitled "Theory and Practice of Karst Dynamic System" which published by China Science Press in 2008. And the Chinese Journal "Carsologica Sinica" from 2008 to 2012.

④ **International cooperation:** The comparative study on karst dynamic system and the carbon cycle between Maocun Village of Guilin and Lost River of Kentucky, have been carried out, in the meanwhile, the standards of the typical catchment selection, establishment of monitoring station and calculation of carbon flux for the karst process and carbon sink effect are being made. These standards will be useful to spread this research worldwide.

国际岩溶研究中心自成立以来，一直关注岩溶动力系统的研究，在全球范围积极推广现代岩溶研究新思路。主要表现在以下4个方面：

①**在科研方面**，国际岩溶研究中心尤其关注了岩溶动力系统与碳循环、石笋对古气候环境记录研究和岩溶动力系统中的水文地质、水环境调查和水资源开发利用。

已经完成和正在执行的，与岩溶作用和碳循环相关的科研项目8项；与石笋对古气候环境记录的项目3项；与水文地质、水环境调查和水资源开发利用的项目8项；与可持续发展相关的项目5项；成功地组织相关专家，奔赴2010年西南大旱、2011年山东大旱、2012年云南大旱，利用前期大比例尺的岩溶水文地质调查资料，结合地球物理探测，完成了抗旱找水任务。

②**宣传和推广方面**，在每年国际岩溶研究中心的年报的扉页均印制岩溶动力系统的定义、结构框图和其功能；在每年国际培训班课程中均安排国际岩溶研究中心学术委员会主任袁道先院士讲解"岩溶动力系统的起源、结构和功能"。

③**出版物方面**，除了积极将岩溶动力系统相关研究成果发表在学术刊物上，同时还专门参与编著了《科学通报》(Chinese Science Bulletin)，2011年9月，56（26），主题：地质作用与碳循环；《人与生物圈》专辑——石漠化，用通俗语言讲解岩溶动力系统与生态环境、石漠化的关系；《中国岩溶》中文刊物的出版；科学出版社出版专著《岩溶动力系统的理论与实践》。

④**国际合作方面**，开展桂林毛村岩溶流域与肯塔基Lost River流域的对比研究，并从岩溶动力系统与碳循环及碳汇效应角度，建立该研究方向的调查、建站、监测、计算的标准，以便于在全球范围推广，该合作项目正在进行中。

Building platform for information exchange, equipments and instruments purchase, and field experimental sites construction
岩溶科技交流信息平台和野外实验设备购置和实验基地建设

With support from the Chinese Government, the great progresses to build platform for exchanging information, buy equipments and instruments, and construct field experimental sites or demonstrations, have been made:

① Information platform: First, the academic committee of IRCK composing 32 members in which 18 members are from foreign countries, was established. Most of them are outstanding experts in karst with prominent education and research. Therefore, the academic committee members are both consultation, comment and policy-making for scientific activities and also the world contact network for IRCK. Furthermore, all the quarterly newsletters and annual reports have been sent to the members. Second, IRCK webpage was constructed and operating well.

② Equipments and instruments: IRCK have purchased a mass spectrometer MAT 253, stalagmites micro-sampler and element analyzers. These equipments will benefit for reconstructing the paleoclimatic and environmental changes with stalagmites in high-resolution, and strive to explore the past extreme weather events and driving forces; A set of micro-organismsanalysers, such as PGR and DGGE instruments for molecular biology and microbiology were also purchased. Organic carbon analyzer and GC-MS that used to reveal the transformation and internal relations between organic and inorganic in karst dynamic system, were also bought. In the meanwhile, the automatic monitoring equipments to reveal the karst dynamic processes sensitive to climate change and environmental change, were purchased. And fluorescence spectrophotometers for tracer test to determine the karst underground watershed boundaries, were also bought.

③ Field experimental sites and demonstrations: With support from the Chinese Government, the programme for IRCK base with 26 ha has been approved by China Development and Reform Commission. Other field experimental sites and demonstrations include Guohua rocky desertification demonstration, Pingguo County, Guangxi; Yaji karst hydrological demonstration, Guilin; Panlong cave drip water monitoring and stalagmite study demonstration, and Maocun karst dynamic process and carbon sink effect demonstration.

在中国政府的支持下，国际岩溶研究中心在建立岩溶科技交流信息平台、购置野外实验设备和建设实验基地方面取得进展，主要体现在以下几个方面：

①科技交流信息平台建设方面，其一，建立了国际岩溶研究中心学术委员会，学术委员会委员由国内外34名委员组成，国外委员18名，都是在国际岩溶学术界学术造诣高，教学科研业绩突出，严谨治学，秉公办事专家、学者，因此学术委员会既是中心学术审议、决策与咨询机构，同时也是国际岩溶研究中心联系世界各国岩溶学术界的网络，中心根据活动情况，每季度撰写季度报、每年撰写年报，均及时分送各位委员；其二，国际岩溶研究中心网页建设。

②实验设备购置，购置了质谱仪MAT253、石笋微区取样器和元素分析仪用于分析石笋对过去气候环境变化高分辨率的分析，力求探索过去极端气候事件的发生及驱动力研究；购置PGR和DGGE等一套微生物和分子生物学分析仪器，组建了岩溶微生物实验室；购置了有机碳分析仪和气质联用仪，用于揭示岩溶动力系统中有机与无机之间的相互转化和内在关系；购置野外动态监测自动记录仪器，捕捉岩溶动力过程对气候环境变化的敏感效应；购置荧光分光光度计，用于示踪试验，确定岩溶地下流域边界。

③实验基地建设，在中国政府的支持下，400亩的国际岩溶研究中心基地项目已经获得中国发展和改革委员会的批准；同时构建了广西平果果化石漠化综合治理研究示范基地、桂林丫吉岩溶水文过程研究示范基地、盘龙洞洞穴滴水监测与石笋中气候替代指标之间关系建立的研究示范基地、桂林毛村岩溶动力过程与岩溶碳汇效应研究示范基地。

Consulting and dissemination activities
科普、咨询与培训活动

IRCK launched a series of training and consulting activities, including:

① **Training courses:** IRCK successfully held five international training courses: karst hydrogeology and karst ecosystem (2009); karst hydrogeology and karst carbon cycle Monitoring (2010); karst hydrogeology investigation technology and methodology (2011); karst and hydrogeology and water chemistry (2012); Karst Hydrogeological Survey, Dynamic Monitoring and Application in River Basins (2013). These training courses attracted 87 trainees from 17 countries and there were 65 lecturers attending our training courses. The training has not only become a worldwide platform for karst international exchanges, it also caused the trainees to be more interested in the research on karst dynamic system. For example, Mr. Mahippong Worakul from Thailand has organized many scientists and managers to visit IRCK and made fruitful scientific exchange, after second training course, when they returned Thailand, they successfully got a large project on karst water resources investigation, stalagmite recording the paleoclimate change and karst process and carbon cycle. Mr. Eko Haryono from Indonesia continuously participated in the training courses in 2009 and 2010, and then became a lecturer in the 2012 training course. He conducted a project on karst process and carbon sink and stalagmite recording paleoclimate change in Indonesia; the trainees from Vietnam, Slovakia, Hungary and Brazil have also made efforts to do relevant study on karst dynamic system.

② **Consulting activities:** As a major consultant of the national project on "The framework on karst rocky desertification comprehensive control (2006-2015)" in China, IRCK has participated in the scientific group on consultation, and then repeatedly accepted the relevant consulting activities. Furthermore, IRCK also established the Guohua Fengcong depression rocky desertification demonstration, Pingguo County, Guangxi. Prof. Yuan Daoxian, the member of the Scientific committee of China National Climate Change, provided useful advices to the committee on the basis of research results from karst dynamic system. Moreover, Sino-US cooperation have been carried out at different levels of training and consulting activities, such as consulting and training activities in Kunming city and Wuming County, Guangxi; as one of the DIKTAS project partners, IRCK provided karst water resource management experience transboundary aquifer between different provinces in China, and also participated in the Fourth Regional Consultation of Asia and Pacific Region on the Groundwater Governance: A Global Framework for Action, in IHP consulting activities.

③ **Dissemination activities:** Fully take the Karst Geology Museum of China as an important site for karst science popularization; as guest editor with "Man and Biosphere" to publish the Special Issue entitled "Rocky Desertification in Karst". To give scientific lectures and make posters in Earth Day and Water Day, in Universities and colleges; actively to participate some dissemination activities in primary schools and high schools, cultivating more students to be interested in karst dynamic system.

国际岩溶研究中心开展了一系列的培训和咨询活动，主要包括：

①培训方面，国际岩溶研究中心成功地围绕岩溶动力系统研究，举办了5期国际培训班：岩溶水文地质与生态（2009）、岩溶水文地质与岩溶碳循环（2010）、岩溶水文地质调查技术方法（2011）、岩溶与水文地球化学（2012）、流域岩溶水文地质调查、动态监测与应用（2013），吸引了17个国家、87位学员参加，同时邀请了65位教员支持培训班。培训班不仅成为在全球范围内推广岩溶动力系统研究方法、岩溶研究国际交流平台，同时，通过培训，也引起相关国家学员对岩溶动力系统研究的兴趣，如泰国学员参加培训班后，回国后，再次组织相关科学家和管理者来中心考察，然后在本国开展了岩溶水资源调查、岩溶石笋和岩溶碳汇等领域的大的科研、调查项目；印度尼西亚学员Eko Haryono，连续参加中心2009年和2010年培训班，到2012年他作为教员参加培训班，同时，在印度尼西亚开展了很好的石笋和岩溶碳汇研究项目；越南、斯洛伐克、匈牙利和巴西等国学员，参加培训班后，纷纷在本国争取科研项目，开展相关的岩溶动力系统研究，使得国际岩溶培训班的成果落实到实处。

②咨询活动方面，国际岩溶研究中心作为中国《岩溶区石漠化综合治理规划大纲(2006-2015)》主要咨询专家，参加大纲的编写，随后多次接受相关的咨询活动，同时建立广西平果果化峰丛洼地石漠化综合治理示范基地；国际岩溶研究中心学术委员会主任袁道先院士，作为中国国家气候变化专家委员会成员，提供来自岩溶动力系统与全球气候变化研究的成果，为中国应对气候变化提供咨询意见；中美合作开展了不同层次的咨询、培训活动，如中国昆明的咨询、培训活动；广西武鸣县咨询、培训活动；作为DIKTAS项目的合作伙伴，提供中国跨省岩溶水资源管理办法案例，作为咨询专家参加IHP的咨询活动等。

③科普活动方面，充分利用中国岩溶地质馆的科普基地作用；与《人与生物圈》杂志合作，用通俗语言出版"石漠化"专辑，达到科普效果；利用地球日、水日开展科普讲座活动；积极参加中小学的相关科普活动，让岩溶动力系统研究的重要性和意义在中小学得到科普，引发年轻一代对岩溶动力系统的兴趣，进而深入探究。

International exchange and cooperation
国际交流与合作

IRCK has implemented extensive international exchange and cooperation in the first six years:

① **Contacts with relevant international organizations:** through participation in the 37th, 38th, and 40th Session of the IGCP Scientific Committee, and submission the quarterly newsletters and annual reports to IGCP Secretariat, IGCP Secretariat can know the activities of IRCK well; In cooperation with the UNESCO Beijing Office to organize IRCK training courses; To participate in the activities organized by China National Committee to UNESCO, IGCP; To visit IHP headquarter in Paris, and to be involved in some consultation, as a DIKTAS partner to give Chinese experience on karst aquifer management, as one of the consultant to take part in the consulting activities in Shijiazhuang, 2012, which contributed to the Groundwater Governance: A Global Framework for Action; Actively to attend the congresses and conferences of IAH, IGC, IGU, IUS and other international conferences, sometimes in charge of the convener sessions related to karst; In cooperation with the Chinese National Committee for MAB to publish Special Issue; In the meanwhile, to invite scientists and experts from different countries as lecturers, such as karst Research Institute of Slovenia, karst institute in MALAGA University of Spain, the National cave and Karst Research Institute in US, and the Institute of Karst hydrogeology in Belgrade University of Serbia, these can enhance partnership and strengthen cooperation.

② **International academic exchange activities:** the "International Symposiumon Karst Water under Global Change Pressure" was held in Guilin, China in April 2013, this International Symposium sponsored by the IAH Karst Committee and China Geological Survey, organized by IRCK/IKG; With International Working Group of IGCP/SIDA 598, three symposia were also held; Actively to attend the congresses and conferences of IAH, IGC, IGU, IUS and other international conferences; To complete 22 receptions from the Ministry of Education and Research of Germany, the Russian Federation Geology and Mineral Resources Department, CCOP, Geosciences and Natural Resources research Institute Germany and other academic organizations; Three international cooperation projects to be carried out.

③ **Management experience exchange and cooperation:** IRCK is the first category II centre concerning geosciences under the framework of UNESCO, and lack of experience in the management. Therefore, IRCK has arranged visits to the Center for Hydrogeology of University of Neuchâtel (CHYN) of Switzerland, the Institute of Water Resources Management Hydrogeology and Geophysics in Graz, Austrian, Queensland University of Australia, and the International Research and Training Center on Erosion and Sedimentation (IRTCES) in Beijing, China. 12 international cooperation MOUs were signed with international organizations and research institutes. And 5 young researchers were sucessfully cultivated to fill the post-doctorial positions, for example, Dr. Li Qiang, as a post-doctoral visiting scholar, went to the University of Mainz, Germany, to carry out the study on the extraction of plant carbonic anhydrase and its stability.

国际岩溶研究中心在第一个 6 年建设期，开展了较为广泛的交流与合作：

①**与相关国际组织保持交往**，通过参加 IGCP 37、38、40 届科学委员会，每季度、每年向 IGCP 秘书处报送中心活动情况，而且保持经常性的活动；与 UNESCO 北京办公室合作举办培训班，及参加教科文组织中国国家委员会、IGCP 中国国家委员会的相关活动，保持着中心与教科文组织间的互动；拜访 IHP、交流岩溶地下水资源的重要性、岩溶区的水害防治及岩溶地下水对全球气候变化的响应，作为 DIKTAS 的合作伙伴，参加石家庄咨询活动等与 IHP 保持经常性活动；参加 IAH、IGC、IGU、IUS 等国际会议，并承担有关岩溶领域专题分会场的召集人；与 MAB 中国国家委员会合作出版有关岩溶石漠化专辑；同时，通过国际培训班，邀请了不同国家岩溶研究优势单位的专家，作为教员，参加培训班，增进了友谊、加强了合作，如斯洛文尼亚岩溶所、西班牙 MALAGA 大学岩溶研究所、巴西岩溶研究所、美国国家洞穴与岩溶研究所、塞尔维亚贝尔格莱德大学岩溶水文地质研究所等。

②**国际学术交流活动**，由 IAH 岩溶专业委员会与中国地质调查局主办、国际岩溶研究中心/岩溶地质研究所成功地举办"International Symposium on Karst Water under Global Change Pressure"国际研讨会，同时与 IGCP/SIDA 598 国际工作组联合举办了 3 次不同岩溶领域的专题研讨会；参加 IAH、IGC、IGU、IUS 等国际会议；完成包括德国教育科研部、俄罗斯联邦地质矿产总署、CCOP、德国地学与自然资源研究院等学术接待 22 批次；开展实质性的国际合作项目 3 项。

③**管理经验交流与合作**，国际岩溶研究中心是教科文组织框架下的第一个地学领域的二类研究中心，在管理方面缺乏经验，先后安排了瑞士纳沙泰尔大学水文地质中心、奥地利格拉茨水资源管理与地球物理研究所、澳大利亚昆士兰大学、国际泥沙研究培训中心交流与访问，取得良好的考察效果；与 12 个国际组织和科研单位签署了合作协议；成功地开展了 5 人次的交流与培训，如李强博士作为博士后访问学者，到德国美因兹大学开展植物碳酸酐酶的提取与稳定性检测方法研究。

Launching a global monitoring network on karst process and carbon sink
启动了全球岩溶动力系统与岩溶碳汇效应监测网络工作

The programme of a global monitoring network on karst process and carbon sink was supported by IRCK Academic Committee and Governing Board in 2011, the progresses of the ongoing programme are:

(1) Comparative research work between Lost River karst underground river basin, Kentucky and Maocun karst underground river basin Guilin, has been done, and a draft report of Standard Operating Procedure for a Global Network of Stations to Measure the CO_2 sink from Carbonate Mineral Weathering is also made;

(2) Slovenia, Thailand, Indonesia, Vietnam have joined and launched the monitoring station work;

(3) Brazil, Hungary, and Slovakia have the intention to join the work.

2011年，在国际岩溶研究中心理事会、学术委员会召开之际，启动了全球岩溶动力系统与岩溶碳汇效应监测网络工作，目前取得的进展包括：

(1) 美国肯塔基州Lost River岩溶地下河流域与中国桂林毛村岩溶地下河的对比研究工作；

(2) 斯洛文尼亚、泰国、印度尼西亚、越南启动了建站工作；

(3) 巴西、匈牙利、斯洛伐克已有意向加入监测网络工作。

The Support of IRCK to UNESCO
国际岩溶研究中心对教科文组织的支持

Purpose of UNESCO
教科文组织的宗旨

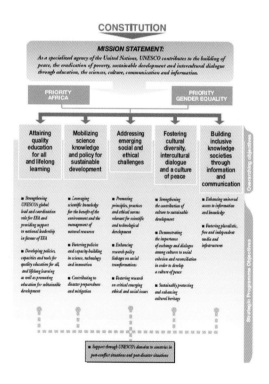

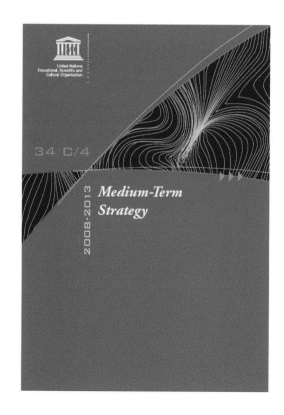

IRCK is the first category II centre concerning geosciences under the auspices of UNESCO, it is under the framework of the International Geoscience Programme (IGCP), UNESCO Natural Sciences. The geoscience of UNESCO includes the sections of IGCP, Earth Science Education in Africa, Earth observation, and Geological parks. And the Natural Sciences Sector contributes to UNESCO's mission by using science to build peace, to eradicate poverty and to promote sustainable development, the priority themes of UNESCO Natural Sciences includes natural resource management, global climate change and adaptation .

UNESCO guild-line document is "Medium-Term Strategy (34C/4, 2008–2013)", in the document it clearly put Africa as the priority region .

Combined with IRCK objectives, the IRCK makes contribution to UNESCO in the following aspects: global climate change, natural resource management, poverty eradication and sustainable development, Earth Science Education (training and fostering) in karst. In IRCK activities, the developing countries, especially in African countries were given more attention.

国际岩溶研究中心是 UNESCO 第一个地学领域二类研究中心，隶属 UNESCO 自然科学部地球科学框架下，地球科学包括国际地球科学计划、非洲地球科学教育、地球观测、地质公园等 4 个部分，而自然科学部的目标是通过科学建立和平、消除贫困、促进可持续发展（The Natural Sciences Sector contributes to UNESCO's mission by using science to build peace to eradicate poverty and to promote sustainable development），自然科学部的优先主题包括了自然资源管理、全球气候变化与适应。

UNESCO 工作纲领性文件是《中期战略（34C/4，2008-2013）》，文件中明确其工作的重点区域是非洲。

因此，结合国际岩溶研究中心本身的特点和目标，国际岩溶研究中心在全球气候变化、自然资源管理、消除贫困与可持续发展及地球科学教育(培训)等方面在岩溶区(岩溶领域)提供支持，在工作开展过程注意对发展中国家，尤其非洲地区的国家倾斜。

Karst dynamic system and global climate change
岩溶与全球气候变化

There are two aspects that make karst dynamic system to keep close relationship with global climate change:

Limestone weathering removal CO_2 from air and global carbon sink effect: the distribution of carbonate rock accounts for 9.3%–15.9 % over the continent land surface. Dissolution of carbonate rocks can remove atmospheric/soil CO_2 to the hydrosphere, and promote photosynthesis of hydrates in aquatic ecosystem, transfer the inorganic carbon into organic carbon, and this is the major part of the carbon flux of DIC transport from land to ocean. The existing estimated data show the carbon flux is 0.1–0.6 PgC/a. If taking the middle value of 0.3 PgC/a, the carbon flux is accounted for 17.65% of that of terrestrial plants, and 37.5% of soil carbon sequestration potential. Considering the deposited organic carbon in terrestrial aquatic ecosystem, the carbon flux from carbonate rocks dissolution will be larger.

Stalagmite can record the paleoclimate change and environmental change: In the study of paleoclimate reconstruction, the ice cores, loess and deep-sea sediments are good recording carriers, when the U–Th dating techniques improved, karst caves stalagmite becomes perfect recording carrier with precise dating and high resolution; stalagmite is expected to reconstruct the past years of extreme weather events; As karst is widely distributed on the global scale, the use of stalagmite records of past climate change can benefit to better understand global climate change.

岩溶与全球气候变化研究主要体现在两个方面：

岩溶动力过程与全球碳汇效应，全球碳酸盐岩的分布占陆地面积的 9.3%–15.9%，碳酸盐岩的溶解过程可以将大气/土壤中 CO_2 转移到水圈中，并促进水生态系统中植被的光合作用，使无机碳转化为有机碳，也是陆地水域向海洋水域输送 DIC 的主体，已有的资料数据显示其碳通量估算为 0.1–0.6 PgC/a，如取中间值 0.3 PgC/a，则该碳通量占到占陆地植物碳汇通量的 17.65%、土壤碳汇潜力的 37.5%，如果考虑陆地水域沉积下来的有机碳通量，这一碳通量会更多。

岩溶洞穴石笋对过去气候变化、环境变化的记录，在对过去气候变化记录研究中，冰心、黄土和深海沉积物是很好的记录载体，而随着 U–Th 测年技术的完善，岩溶洞穴石笋成为测年准确、分辨率高的不可多得的研究载体；石笋载体可望重建过去年际、年间的极端气候事件；鉴于岩溶在全球范围的广泛分布，利用石笋对过去气候变化记录高分辨率的特点，可以帮助人们更好地认识全球气候变化规律。

The utilization and management of water resources in karst area
岩溶区水资源开发利用及管理

Karst hydrogeological structure is characterized with double layers, when the rainwater falls onto surface quickly through epikarst zone, vadose, vertical migration removal to underground rivers. Therefore, water resources in karst area show lack of surface water, but abundant ground water. Globally about 25 percent of the world's population rely on the karst water, it has an important strategic position.

IRCK/IKG has been focusing on the karst hydrogeological investigations, karst groundwater exploitation, utilization and protection. The Chinese Government has also given a lot of support, and karst hydrogeological investigations, karst underground river survey and demonstration of karst groundwater resources exploitation and utilization have been carried out, these work approach and reach a good results:

(1) Hydrogeologic survey with 150,000 has been done to 200,000 square kilometers, this provided the essential data and maps to fight the extreme drought during 2009-2012, organized by the China Ministry of Land and Resources, and roughly 80% successful drill-well happened in karst area;

(2) More than 3,000 underground rivers and more than 60 kilometers caving have been investigated. The existing data show the total length of the underground river is more than 14,000 km and the recharge area with approximately 300,000 km^2, and the runoff of the ground rivers in dry season is up to 47 billion m^3/a, equivalent to the total runoff in the Yellow River;

(3) Depending on karst hydrogeological background, many models for karst groundwater exploitation and utilization have been produced. And these provide drinking water and farmland irrigation water for the local residents;

(4) Initially to pay more attention on the water environment: Preliminary investigation on karst groundwater contamination, and analysis the sources of groundwater contaminants, exploration the control measures on underground river pollution.

岩溶水文地质结构以地表、地下双层结构为特征，地表降水通过岩溶表层带、包气带垂向运移至饱水带和地下河。因此，从水资源的空间分布情况来看，岩溶区十分缺乏地表水系，但地下水系发育。从全球尺度看，岩溶水为全球25%的人口提供生活、生产用水，尤其饮用水，具有重要的战略地位。

国际岩溶研究中心/岩溶地质研究所一直将岩溶水文地质调查、岩溶地下水开发、利用和保护，作为工作的重点，中国政府也给予了很大的支持，开展了不同尺度的岩溶水文地质调查、岩溶地下河调查、岩溶地下水资源开发利用示范等工作，取得了较好的成果：

(1) 完成1：5万水文地质调查面积20万 km^2，为2009-2012年国土资源部组织岩溶区抗旱找水打井行动提供了基础数据；

(2) 调查统计了3000多条地下河，探测岩溶洞穴60 km，地下河总长度超过1.4万 km，汇水面积约达30万 km^2，枯水季节径流量也高达470亿 m^3/a，相当于一条黄河的径流量；

(3) 根据岩溶水文地质条件，形成多个岩溶地下水开发、利用模式，支持了当地的人畜饮水和耕地灌溉问题；

(4) 初步调查了西南岩溶区地下水的污染状况，分析了地下水污染的来源，探索了地下河污染的防治措施。

Poverty elimination and sustainable development in karst areas
岩溶区消除贫困与可持续发展

In order to improve the ecological environment and eliminate poverty, IRCK mainly has carried out the following work:

(1) Rocky desertification Control. Southwest is a major karst area in China and also a concentrated poor population. In order to eliminate poverty and enhance ecological restoration, the Chinese central government launched a national engineering programme "Integrating measure for rock desertification in karst area" in 2006. The four strategies for rocky desertification control are being put forward: ① water is the front problem; ② soil is the key problem; ③ the plants, especially economics plants species selection is very important; ④ the harmony development between the ecology and economics, and local people to eliminate poverty is the goal. In order to better promote the the project, IRCK also has made summaries models to rocky desertification control in different karst geology and one demonstration in Fengcong karst area.

(2) Soil loss control. According to the process of soil erosion in karst areas, propose different soil and water conservation measures: ① Closing the upper area on the slope and fencing off for afforestation; ② In the middle-lower slope, making bench terrace to protect soil, and cultivating high economic yield and drought-resistant crops; ③ In the depression and doline, engineering measures to prevent soil to go into the underground river and fighting the flood; ④ Restoring vegetation on upstream, farming area protection, and improving soil and water resources utilization efficiency in discharge zone.

(3) Karst collapse is a common geological disaster in karst areas. At present, a set of "formation mechanism model, risk assessment and mapping, monitoring and forecasting and information management" to karst collapse prevention system and method have been developed.

(4) IRCK actively participated in the consultancy activities, preparation the proposals and management strategies for karst geopark and World Natural Heritage, including Geopark of Shuanghe Cave in Suiyang, Guizhou; Geopark of Leye, Fengshan, and Xiangqiao in Guangxi, and Geopark of Wulong in Chongqing.

围绕改善生态环境和消除贫困人口，国际岩溶研究中心主要开展了以下工作：

(1) 石漠化治理。中国西南是主要岩溶分布区，同时也是中国贫困人口集中分布区，为了消除贫困、恢复生态，中国政府在2006年启动《岩溶区石漠化综合治理工程项目》，其综合治理的对策可归纳为：水是龙头，土是关键，植被(经济植物)是根本，区域生态经济双赢、农民脱贫致富是目标。同时为了更好地推进该工程项目的顺利进展，国际岩溶研究中心还总结了不同岩溶环境下石漠化治理示范模式。

(2) 水土流失防治。根据岩溶区水土流失的过程和特点，提出水土保持对策：坡地上部封山育林、栽种有水保效益、根系发达的灌木；坡地的中下部砌墙保土、栽培经济效益高、耐旱的经济作物；洼地底部防止土壤随水流进入地下河、季节性涝灾；地下河流域的上游封山育林、恢复植被，径流区保护性的农耕，排泄区提高水土资源利用效率。

(3) 岩溶塌陷是岩溶区常见的地质灾害。目前，已初步形成一套以"形成机理的模型试验、风险评估与制图、监测预报和信息管理"为特色的岩溶灾害防治研究体系与方法。

(4) 积极参与岩溶区地质公园、世界自然遗产地申报和管理咨询活动。主要包括：贵州绥阳双河洞，广西乐业、凤山、香桥，重庆武隆等处。

Education of karst geology (training)
岩溶地质教育（培训）

Karst geology education of IRCK in a variety of approaches:

(1) To fully play the role of the Karst Geology Museum of China in dissemination, to make the lectures on karst science in universities and colleges, and to organize the Earth Day and Water Day;

(2) With personnel exchanges, to train young generation to understand the karst dynamic system well and to learn modern technologies, these are beneficial to multidisciplinary and integrating study;

(3) Through annual training, to explain the karst dynamic system, structural characteristics and its applications, more and more trainees from different countries were cultivated on modern karst. So far, there were 87 trainees from 17 countries attended IRCK's training courses.

(4) Postgraduate students have been trained: IRCK cultivated 15 PhD students and 50 master students during 2008–2013, their disciplines are karst geochemistry, karst hydrogeology, karst environment, karst ecosystems and karst geological engineering.

国际岩溶研究中心通过多种渠道开展形式多样的岩溶地质教育活动：

(1) 充分利用中国岩溶地质馆的科普基地作用，岩溶科学走进大学，地球日、水日等开展岩溶科学的宣传、教育活动；

(2) 通过人才交流，培养年轻人，认识和掌握岩溶动力系统和研究方法，掌握现代技术手段，推进不同学科间的交叉渗透；

(3) 通过每年的培训，系统讲解岩溶动力系统运行规律、结构特征及其应用，包括2013年第五次国际培训班，共培训来自17个国家的87位学员。

(4) 研究生、博士生的培养，依托国际岩溶研究中心导师培养的博士15人、硕士50人，专业包括岩溶地球化学、岩溶水文地质、岩溶环境学、岩溶生态系统和岩溶地质工程等。

Incline towards developing countries, especially African countries
对发展中国家,尤其非洲地区的国家倾斜

In line with UNESCO's priority area, IRCK's work focuses on developing countries, especially African countries.

(1) The trainees of IRCK's training courses mainly from developing countries and African countries, so far, 85 participants from developing countries, accounting for 98%, of which 22 from African countries, accounting for 25%;

(2) A cooperation agreement between the University of the Western Cape with IRCK was signed in 2011 and the China-Africa Water Resources Forum has been formed, and the first meeting was held in Cape Town, July 2013. The Forum aims to face the challenges on problems of hydrology, resources and environment, under the global climate change, rapid urbanization, to strengthen both China and African countries exchanges and cooperation, to explore effective counter measures to the sustainable utilization of water resources.

根据教科文组织工作部署的重点区域,国际岩溶研究中心在工作开展过程注意对发展中国家,尤其非洲地区的国家倾斜。主要体现在以下两个方面:

(1) 在国际培训班学员招生中,5 次培训班,学员总数 87 名,其中来自发展中国家的学员 85 名,占 98%,其中来自非洲国家的学员 22 名,占 25%;

(2) 与西开普大学、教科文组织水文地质教席签署了合作协议,并促成了中非水资源论坛,2013 年 7 月在开普敦召开了第一次会议,该论坛的宗旨是面临全球气候变化、快速的城镇化发展带来的水文问题、资源环境挑战,加强中国与非洲国家之间的交流合作,共商有效对策,以期水资源的可持续利用。

Postscript

The International Research Center on Karst (IRCK) under the Auspices of UNESCO has gone through the first six-year journey, and its achievements were approved by the Evaluation Panel appointed by UNESCO. The achievements stands for that IRCK has completed the first phase of the establishment and task.

During the establishment and development of IRCK, tremendous supports were received from experts and scholars from many institutes nationally and internationally. These include UNESCO-Geology, Ecosystems and Biodiversity, UNESCO-IHE Institute for Water Education, The Chinese Government, The National Development and Reform Commission, Ministry of Science and Technology, National Natural Science Foundation of China, Ministry of Land and Resources, China Geological Survey, Chinese Academy of Geological Sciences, Institute of Karst Geology and other domestic and foreign institutes and organizations. On the occasion of the publication of this book, we would like to give our sincere thanks for their selfless contributions.

IRCK has entered into the second six-year construction period, and it needs to make greater progress on the basis of the existing foundation and experience in karst technological innovation. Two aspects should be focused on in future: First, IRCK continues the study on the relationship between karst dynamic system and global climate change. In 2013, IRCK which focuses on the karst dynamic system and global change was approved by China Ministry of Science and Technology to be a national and international joint research center. The purpose of the center is to address the global climate change using karst disciplines. IRCK provides an active academic platform to ally the national and international institutes to promote and carry out related research work. However, IRCK is also requested to follow the relevant regulations formulated by certain international research institutes. Second, IRCK provides training and advisory service in karst research areas for the developing countries; specifically, on the topics of water resources, environmental and ecological issues. IRCK aims to achieve the goal entitled "purpose of earth sciences for the service of humanity" proposed by UNESCO—Geology, Ecosystems and Biodiversity.

In the end, we sincerely hope that with the sponsorships from the UNESCO and The Chinese Government, and the supports from National Development and Reform Commission, Ministry of Science and Technology, National Natural Science Foundation of China, Ministry of Land and Resources, China Geological Survey, Chinese Academy of Geological Sciences, Institute of Karst Geology and other domestic and foreign institutes and organizations, IRCK will continue to make great achievements through international exchange and cooperation with karst experts and scholars.

后记

联合国教科文组织国际岩溶研究中心走过了六年的历程，顺利通过了教科文组织评估专家组的评估，标志着国际岩溶研究中心完成了第一期的建立、建设任务。

在国际岩溶研究中心成立和建设过程中，得到教科文组织地学生态部、水科学部，中华人民共和国政府、国家发展和改革委员会、科学技术部、国家自然科学基金委员会、国土资源部、中国地质调查局、中国地质科学院、岩溶地质研究所及国内外相关机构、组织的大力支持，国内外关注、关心国际岩溶研究中心成长的专家、学者的无私奉献，在此仅以此图册向他们表示衷心的感谢。

国际岩溶研究中心进入第二个六年建设期，需要我们基于已有的基础和经验，在岩溶科技创新方面取得更大的进展，重点关注两方面：其一，岩溶动力系统与全球气候变化，2013年，中国科学技术部认证了"岩溶动力系统与全球变化"为国家国际联合研究中心，该中心的宗旨就是应对全球气候变化这个世界性的课题中做出岩溶学科领域的贡献，这就需要我们遵循国际研究机构的相关管理条例，发挥国际岩溶研究中心的平台作用，联合国际优势力量，推动和开展相关科研工作，取得创新性成果；其二，为发展中国家，尤其岩溶区的资源、环境和生态问题提供更多培训、咨询服务，实现教科文组织地学生态部提出的"地球科学为人类服务"的宗旨。

最后，祝愿国际岩溶研究中心，在教科文组织、中华人民共和国政府赞助下，在国家发展和改革委员会、科学技术部、国家自然科学基金委员会、国土资源部、中国地质调查局、中国地质科学院、岩溶地质研究所及国内外相关机构、组织的大力支持下，加强与国内外关注、关心国际岩溶研究中心成长的专家、学者的联系，广泛开展国际交流与合作，团结协作，共同努力，取得新的成绩。